DESIGN FOR FLYING

DESIGN FOR FLYING

DAVID B. THURSTON

Consulting Aeronautical Engineer
President, Thurston Aeromarine Corporation

McGRAW-HILL BOOK COMPANY

New York   St. Louis   San Francisco   Auckland   Bogotá   Düsseldorf   Johannesburg
London   Madrid   Mexico   Montreal   New Delhi   Panama
Paris   São Paulo   Singapore   Sydney   Tokyo   Toronto

RAMSEY • *Budget Flying (1981)*
SACCHI • *Ocean Flying (1979)*
SMITH • *Aircraft Piston Engines (1981)*
THOMAS • *Personal Aircraft Maintenance (1981)*
THURSTON • *Design for Flying (1978)*
THURSTON • *Design for Safety (1980)*
THURSTON • *Home-Built Aircraft (1981)*

Library of Congress Cataloging in Publication Data

Thurston, David B
Design for flying.

Includes index.
1. Airplanes—Design and construction.
I. Title.
TL671.2.T57   629.134'1   77-7384
ISBN 0-07-064553-1

3 4 5 6 7 8 9 0 VHVH   8 3 2 1

The editors for this book were Jeremy Robinson, Gretlyn K.
Blau, and Margaret Lamb, the designer was Elliot Epstein,
and the production supervisor was Teresa F. Leaden. It
was set in Optima by Progressive Typographers.

Printed and bound by Von Hoffmann Press, Inc.

*Frontispiece: Rockwell Commander 112TCA. Rockwell
International-General Aviation Division Photo.*

*To my wife, Evelyn, who for over thirty-five years
has encouraged and endured my interest in aviation.*

CONTENTS

pect ratio vs. climb · drag vs. climb · temperature and altitude vs. climb · stability in climb.

PREFACE

Design for Flying was written in response to many inquiries and questions received over the years concerning aircraft handling characteristics, operational procedures, and detail design. Since the answers are of great interest to a broad cross section of the flying public, this book has been prepared with a minimum of mathematics—with nothing beyond simple algebra required to understand the most complex of the few formulas presented.

Important design criteria are reviewed by running through a flight profile from takeoff to landing, with the critical requirements for each phase of flight presented at three levels: (1) why aircraft must fly as they do to satisfy pilot and FAA handling requirements; (2) what design features make the airplane behave as required; and (3) what design criteria are necessary to establish these characteristics. The first two levels will be of interest to everyone concerned with aircraft, while the technical explanation may be skipped if considered outside any reader's interest. By exploring the relationship between aircraft design features and flight characteristics, *Design for Flying* invites every reader to examine the effects of his or her own piloting technique upon aircraft response.

As a matter of interest and to provide additional background information, the history and development of essential criteria are briefly reviewed along with the various phases of design. Although few basically new concepts have been introduced into the aircraft field since 1911, the old ideas are modernized from time to time and, where relevant, their effect upon current aircraft is noted. For readers interested in digging further into technical items or designing their own airplane, useful design and maintenance references follow each chapter.

Further, if nontechnical readers wish to change the speed or climb range of their propeller, equip their airplane for IFR flight, position avionics antennas, or modify existing equipment items, *Design for Flying* tells what must be done and how to go about doing it—information that should save every aircraft owner both time and money.

As a result, this book really is a general aviation encyclopedia providing an excellent background and reference source for ground and flight school instruction, technical schools and individuals interested in practical design, and all pilots curious about how aircraft are designed and what makes them fly as they do.

Naturally, during 35 plus years of active design experience many people have influenced my interest in the aircraft field. In addition to expressing my sincere appreciation to everyone of this great number, I would particularly like to thank the following collaborators who were of assistance in getting this book together:

The publishers and editors of *Aero Magazine,* first Marvin Patchen and then Will Hunt, who over the past few years have encouraged and supported my interest in writing and publishing semitechnical aviation articles, some of which have been reedited with their approval to form the basis for parts of this book.

Two capable airmen, Richard A. Bradley and Walter H. Prentice, both of whom are competent Certified Ground and Flight Instructors holding ratings for commercial and instrument training. In addition, Dick is a corporate jet captain and Walt is a qualified aerobatic instructor. Their comments and guidance have made *Design for Flying* a more useful and practical book.

My friend for many, many years, a former test pilot and fellow engineer, Hank (Franklin T.) Kurt, who encouraged preparation of this book and offered many helpful suggestions during its progress.

The many aircraft sales and public relations people who responded rapidly and enthusiastically to requests for photographs and other product information.

And last, but by no mean least, my wife, Evelyn, who had the perseverance to translate my handwriting into the final typed manuscript.

David B. Thurston

1 BASIC DESIGN CONSIDERATIONS

BEGINNING FLIGHT   Ever since humans learned to stand erect they have been preoccupied with a desire to leave the ground and fly in a controlled direction. If we assume that surviving legends, wall decorations, and tapestry details have some basis in fact, humanity's early urge to fly was a small but constant contributor to population reduction.

Since flight is the visible result of many phases of applied physics, positive advancement must be keyed to state-of-the-art engineering, mechanical design, and fabrication techniques. When we review the many early and unsuccessful attempts at flight, it becomes obvious that controlled flight with wings was dependent upon some theory of lift of inclined planes (wings) and so had to await the required mathematical equations. As a result, balloons—air-displacing vehicles—logically provided the first real demonstration of atmospheric flight. As is well known, the Montgolfier brothers, French manufacturers of fine paper, are generally credited with the first public balloon flight on September 19, 1783, using hot air for lift. However, 3 weeks earlier the distinguished French physicist Jacques Charles (discoverer of Charles's law) had secretly launched an unmanned hydrogen-filled balloon which flew for some distance to the terror of all local peasantry.

Typical of aeronautical development, a field in which any advance is eagerly pursued by all participants, balloons rapidly proliferated everywhere. I am always amazed at the amount of time, money, and enthusiasm people seem to find today for new development projects of every sort, and apparently this was just as true in the late eighteenth century. Balloon flights soon crossed the English Channel and the Alps, thrilled crowds in North and South America, and naturally, were applied at once to military observation—where balloons were still successfuly employed as recently as World War II.

Although many fortunes were spent attempting winged flight, nothing really new developed until a text covering the physics of fluid motion was published by Horace Lamb in 1879. Combining this work with existing theories such as Bernoulli's principle finally permitted rational calculation of the lift and center of pressure of flat and cambered surfaces. In 1885 Prof. John J. Montgomery of Santa Clara, California, successfully flew a manned glider with aileron control as shown in Fig. 1-1, subsequently publishing his equations for lift of flat and curved planes in 1890. Many other mathematicians were at work during this same period, with the result that we were scientifically more or less ready for powered flight by 1900 but lacked an adequate engine.

Every quantum jump in progress requires some fundamental technical advancement. Just as manned space flight awaited development of the solid-state computer, manned atmospheric flight was dependent upon an efficient powerplant of some sort. This contribution was made by the Wright brothers through development of a gasoline engine and drive system superior to any then available.

Certainly another fundamental achievement made by the Wrights was their technical analysis of aerodynamic design before starting full-scale construction and flight demonstration. Most of the early pioneers were trial-and-error specialists and the law of survival of the fittest surely applied; as a result any single flight completed without injury or fatality contributed to aeronautical knowledge. Fortunately, because of the many participants, aircraft and engine design and performance steadily improved through development by ordeal.

The importance of lateral control should be briefly reviewed at this point not only to provide background information but also to emphasize that lateral control is very basic to successful flight; it is a critical control for crosswind opera-

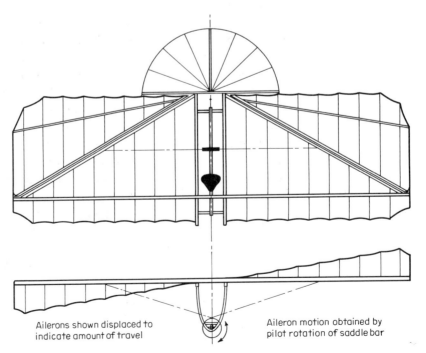

FIGURE **1-1**
Montgomery's aileron-
equipped glider of 1885

Ailerons shown displaced to
indicate amount of travel

Aileron motion obtained by
pilot rotation of saddle bar

tion near the ground and also essential for all coordinated flight maneuvers. With the exception of early designs patented by Montgomery of the United States and Louis Pierre Mouillard of France there is no evidence that the early experimenters were aware of the need for lateral control. In fact, most designers were apparently so involved in the basic problems of becoming airborne that any requirement for a controlled turn was secondary to getting off the ground for a measurable period of time. And yet lateral control was to become a problem as basic as that of structural design and efficient power. Every current competition for man-powered flight requires completion of a figure-eight flight path, providing recognition of the basic difficulty of controlled turning flight.

The fact that the Wrights and Glenn Curtiss did realize the fundamental need for lateral control partially explains their early success. Using a different lateral-control approach from the Wright's wing warping system, Curtiss developed separate ailerons initially hinged about the wing leading edge. These ailerons, located at the tip of each wing, were successfully demonstrated on the 1908 June Bug and were subsequently used on Curtiss's designs in the arrangement shown on Fig. 1-2.

Similarly, until the Wright brothers installed rudder control in late 1904 little thought had been given to use of a vertical surface for turn coordination and assistance. As a matter of interest, the first recorded circular flight under power was made by Wilbur Wright on September 20, 1904, using wing warping plus rudder control. But the added drag of turning flight combined with marginal thrust and poor pilot technique resulted in many crashes for many pilots before both power and skill were developed to a level permitting practical and safe controlled flight.

Despite all obstacles, the art of powered flight developed very rapidly, and by 1911 aircraft of all configurations had been built and flown. In fact, with the exception of fully retractable landing gear, I do not believe a basically new arrangement of wings, tail surfaces, control system, alighting gear (land or water), or reciprocating powerplant has been developed since 1911. By that date Curtiss had flown tricycle landing gear, developed a wheel control with integral throttle, produced relatively high-powered/low-weight engines, and designed amphibious floats with externally retractable wheels. In addition to correctly analyzing the criteria for a coordinated turn and developing the required aerodynamic configuration, the Wright brothers had originated the skid landing gear and the catapult launching system (similar to present aircraft carrier practice), improved engine reliability and propeller efficiency, and refined their structural design procedures sufficiently to satisfy basic military procurement requirements.

By 1911 other powered aircraft experimenters had developed tandem and four-wheeled landing gear as well as the tail-wheel-type; a tractor monoplane had flown across the English Channel destroying forever the isolation of the British Isles; Paul Cornu had designed and built a model V/STOL (vertical/short takeoff and landing) helicopter having a wing, rotors that could be deflected to permit forward flight, and a skid landing gear; propellers were located at the fuselage nose and tail as well as on aircraft wings; canard, tandem, monoplane, biplane, and triplane wing arrangements had been flown with varying degrees of success; wheel, stick, shoulder yoke, and rudder bar were used in arbitrary combinations for turning and climbing flight; and rotary and reciprocating air-cooled and water-cooled gasoline engines had been flown, as had wooden and metal-bladed fixed and variable pitch propellers.

According to the 1910 text *Aerial Navigation of To-Day,* Thomas Edison had recently predicted the helicopter would be the flying machine of the future, designers considered the monoplane superior to the biplane for high-speed flight, Russia had stated her right to fire at any balloon crossing her frontier, and European officials were concerned about possible "smuggling by aeronauts." So not much has happened in the past 7 decades of aeronautical development that was not tested or anticipated by 1911.

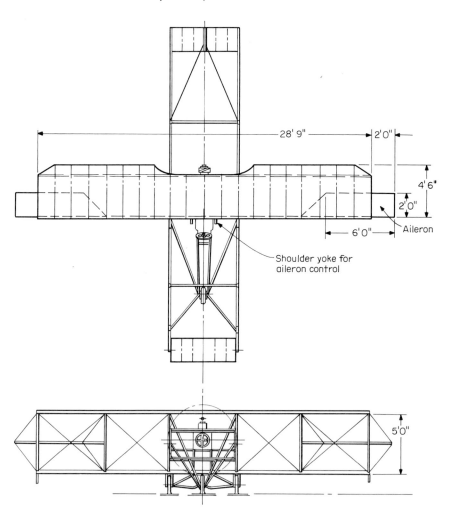

FIGURE **1-2**
Curtiss 1910 biplane showing
the mid-gap ailerons and
shoulder yoke control

THE EVOLUTION OF DESIGN SPECIFICATIONS

The improvement of aircraft design from 1911 concepts to current levels has paralleled the combination of practical experience gained from flight plus the refinement of engineering techniques realized from structural test, flight test, and materials development programs.

Despite the flight evaluation of a multitude of design configurations by 1911, many of these aircraft would have been extremely dangerous to fly if adequate power had been available. Airframe structural design was also an unknown art conducted largely by trial and error, with frequently fatal results due to in-flight structural failures. Further, there is no indication that any early designer considered the existence of gusts or their effect upon flight loads and flight-path control during the approach. This gap in essential background analysis developed primarily because flight was a new art which every experimenter and adventurer wished to savor immediately; rational solutions simply could not be realized within the time and funding limits available. In many ways the aviation period from 1908 through 1912 was similar to the California Gold Rush of 1849; everyone wanted to get into the act and by any route possible.

Due to poor evaluation of flight requirements, many stall and skid crashes were never explained or fully understood during this early development period. However, through data obtained from a number of such accidents the Army Signal Corps gradually evolved an aircraft design procurement specification, with similar requirements established by the Navy for its air arm.

As would be expected, military procurement was the principal support of the struggling aircraft industry through World War I, with accompanying design specifications growing in length and detail as more was learned about aerodynamics and high strength-to-weight ratio structural design. When the fighting was over, so was the first peak of aviation growth. The marketplace was flooded with surplus military aircraft, notably the Curtiss Jenny, the DH-4, and similar aircraft that could be used by sports enthusiasts and barnstorming pilots; a sales pattern unfortunately repeated after World War II. When vacant fields of

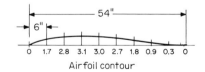

Airfoil contour

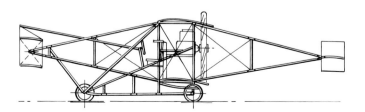

reasonable size started to sprout tandem-seated biplanes all across the United States, the Department of Commerce was appointed as licensing agent for aircraft and pilots, with authority to draft basic requirements for commercial aviation in the United States—and so established the foundation for our present Federal Aviation Administration (FAA).

During the period between 1920 and 1929 many prize award flights were undertaken both here and abroad, some between great distances—viz. Lindbergh's May 1927 transatlantic flight from Roosevelt Field to Paris. The airframe, engine, and instrument design requirements specified by the competing pilots and financing syndicates effected a major improvement in aircraft quality and reliability. Stability became of increasing importance during record flights, while the night mail runs indicated the great need for blind flight instrumentation and icing protection in addition to inherent stability.

Whenever flight evaluation demonstrated that a design modification, new device, or flight performance requirement directly contributed to improved flight safety, it would eventually be incorporated into the Airworthiness Requirements. These specifications were controlled by the Bureau of Air Commerce through the thirties, then the Civil Aeronautics Administration, the Federal Aviation Authority, and now the Federal Aviation Administration acting as an arm of the Department of Transportation. And so we can see that aircraft design regulations developed from early Signal Corps days to the present through a rather obvious but natural Darwinian process of evolution.

LIGHTPLANE DESIGN RESPONSIBILITY

Unless they are in the unusual position of having a production order for a specific airplane design or are employed by one of the "big three" private aircraft manufacturers and so have at least a dealer pipeline to fill, designers of new lightplanes may find no market for their products. In that case, prototype development time and money have been unfortunately wasted, and the sponsoring company may be forced out of business. This scenario has been repeated all too frequently in the lightplane business and explains why bankers take a basically dim view of financing new ventures in the private aircraft field. It also partially explains why manufacturers are hesitant to introduce radical designs, a subject reviewed in some detail in Chap. 3.

The successful company must offer the market an airplane of sound structural and mechanical design plus quality construction. The designer's and the manufacturer's reputations are probably more closely related to product integrity in the aircraft industry than in any other field. The explanation is quite fundamental: people's lives are at stake every time the product is used. An easily handled, good performing, structurally sound airplane is rapidly accepted and will establish the reputation of the design and manufacturing company. Good news travels fast in the aircraft industry since a pilot in New England today may be in Tennessee tomorrow and Arizona the day after. Of course, bad news travels at the same speed and seems to grow with time and distance as well. So designers must be constantly aware of their responsibility to produce a sound, useful airplane—which may frequently have to exceed FAA regulations in order to satisfy their personal criteria for safe operation and low maintenance. Let us see how these goals may be realized.

In order to improve operational safety the designer must eliminate as may features as possible that could lead to pilot error in flight or on the ground. To achieve this basically requires:

· An aerodynamic arrangement offering ample stall warning through buffeting and possibly "soft controls" which are apparent 5 to 10 mph above the stall (soft controls means a noticeable reduction in control forces)

· A stable airplane that can be readily trimmed during all phases of flight operation

· A rate of climb exceeding 700 ft/min

· A well-designed and tested structure

· A proven, reliable powerplant

· An automatic alternate source for engine intake air

· Simple, "fool proof" engine-starting and fuel-delivery systems

· Excellent pilot visibility during taxi and flight

· A relatively quiet cabin to reduce pilot fatigue during long trips

This is not a tall order today; such features combined with modern radio navigation represent virtually the entire advance in general aviation design since 1922. Our average personal aircraft really does not fly any faster or much more efficiently today, but we can comfortably navigate from A to B in relative safety, with a reliable engine, and in an easily handled airplane.

Although flight safety must be considered of paramount importance for obvious reasons, designing for low maintenance is also a prime requirement for a successful airplane. There is nothing more frustrating to an owner, or damaging to the industry, than an airplane that becomes a vast sinkhole for hard-earned (after-tax) dollars. The designer can be of great help here by:

· Providing an efficient structure, one that is rugged without unduly penalizing useful load

· Eliminating complex hydraulic and electrical systems by using mechanical operation of flaps and landing gear

· Developing a clean design that obtains high cruise with fixed, faired landing gear

· Locating batteries, relays, brake system reservoirs, pitot lines, etc., in accessible areas

· Providing an instrument panel that is readily removable for service or inspection

· Using circuit breakers instead of fuses

· Using proven components and accessories that are backed by reputable companies

These maintenance items may seem obvious; yet have you ever tried to find the connection to your vacuum system air filter, fill your battery, or find that hydraulic system slow leak that keeps staining the belly skins? Over a period of time poor detail design can cost much more than the initial purchase price of any airplane.

We will discuss these features in some detail under the related chapters with the hope that many owners will be able to perform more of their own aircraft maintenance and clean-up work within permissible FAA service limits. For example, flight test data show that both speed and rate of climb can be increased by putting wheel fairings on a fixed tricycle landing gear. While some gain would be expected, the level of improvement was found to approach that of a fully retractable landing gear. A fixed landing gear means lower initial cost and maintenance plus one less opportunity for pilot error, since you can't land gear-up if the gear doesn't retract. These are the reasons most small aircraft have fixed gear equipped with wheel fairings.

THE MOST POPULAR
AIRCRAFT

As an introduction to our study of why flight requirements actually design the successful airplane, we should consider just how a few basic design features relate to modern single-engine aircraft.

If we measure success by ownership popularity, the winner by a factor of at least 2 is the Cessna line of single-engine, fixed gear, four-place airplanes. Typical of this series is the Cessna 172 Skyhawk of Fig. 1-3.

Over 25,000 of these aircraft had been built through 1976. This design has ample stall warning and probably is the most stable of any airplane in its weight and speed class. In fact, I believe the solid feel of the single-engine Cessna models is a major reason for their popularity; they are stable flight platforms and so inspire pilot confidence.

Why is this so? To be stable in hands-off trimmed flight, which is referred to as *stick free stability,* an airplane must have a fairly long tail arm (which is the distance from the center of gravity to the center of lift of the horizontal or vertical tail). The horizontal tail should also be rather large to provide the rapid damping necessary to cancel pitch oscillations caused by turbulent air or pilot action. Cessnas have these features as shown by Fig. 1-3.

Responsive controls are necessary at the stall, during approach, and for operation under moderate to strong crosswind conditions. This is particularly true of the ailerons, so plenty of aileron area is required; while tapered outer

FIGURE **1-3**
Cessna 172 Skyhawk
Cessna Aircraft Company Photo

wing panels should be used to reduce roll damping and thereby increase rate of roll for improved lateral control. Again, we find these features on the Cessna models.

The high-wing configuration provides excellent downward visibility in flight, although side vision tends to be restricted in a turn; the accepted technique to overcome this condition is to raise the wing on the intended turn side and look around before starting to bank. However, this potential blind spot is annoying in terminal areas and has undoubtedly been responsible for many "in-pattern" accidents. The larger 180-hp Cessna Cardinal has much improved pilot visibility in turns plus much more elbow room and rate of climb, resulting in a generally superior airplane compared with the 172 Skyhawk—but has a superior price as well. At any rate, most owners must prefer downward visibility over turning visibility because Cessna single-engine sales continue to grow year after year.

Tricycle landing gear is much easier to handle on the average runway than tail-wheel-type gear, especially in crosswind conditions, so modern light aircraft should be expected to have tricycle gear with a steerable nosewheel— and all major production aircraft do have this feature.

By way of satisfying our maintenance items, the Cessna 172 has fixed, faired landing gear and no hydraulic system other than the main wheel brakes. The fixed pitch propeller simplifies flight operation since there is no pitch control requiring pilot attention, while the Lycoming 160-hp (172 Skyhawk) and 180-hp (Cardinal) engines are proven workhorses of record with long hours of operation between major overhauls. Due to the low position of the fuselage floor when parked, the Cessna 172 and Cardinal are both easy to enter and service.

FIGURE 1-4
1976 Piper Cherokee Cruiser
Piper Aircraft Photo

The only real drawback found with the Cessna high-wing configuration occurs during preflight inspection of the fuel tanks located high above ground in the wing panels. However, this is where the fuel should be and external steps can be obtained as an option to ease the fuel inspection problem.

Although the actual contribution of active sales and service departments cannot be fully assessed in relation to total sales volume, it is quite evident that the Cessna line of single-engine personal aircraft satisfy most of the design and maintenance criteria—and Cessna has the sales figures to prove it.

The number-two production position belongs to the Piper Cherokee line of low-wing, tricycle gear aircraft shown in Fig. 1-4. This design has a slightly shorter tail arm than the Cessna 172. It also has an all-movable horizontal tail which can be subject to annoying minor trim changes, apparently caused by local variations in airflow as well as by slight wear and play in the control system linkage. The combination of shorter tail arm and all-movable tail result in somewhat reduced, although adequate, longitudinal stability for the Cherokee series compared with the Cessna fixed stabilizer design. While the basic Cherokee rectangular wing tends to reduce roll rate and crosswind response, the later Piper Warrior and Archer II designs have tapered outer panels which improve aileron effectiveness to meet or exceed Cessna levels.

The Cherokee low-wing arrangement provides two features everyone admires: visibility in the pattern and ease of wing fuel tank inspection. The low wing also offers a convenient structure for mounting the main gear, while the Piper oleo struts are better shock absorbers than the Cessna spring steel main gear legs. The low wing and related support structure also combine to offer slightly superior crash protection during off-field landings, although a study of such accidents indicates the difference in fatalities favoring low-wing aircraft is much lower than might be expected.

Despite the many favorable aspects of a low-wing location, all such designs share a common problem: difficulty of cabin entry. Unless a sliding canopy is used, climbing through a side door positioned above a wing is difficult for anyone who is not a natural acrobat. And even the canopy approach may not be appreciated by women in skirts. However, the Cherokee cabin is roomier than that of the Cessna 172, making up for the more difficult entry.

The basic structure of the Cherokee models, particularly at the wing to fuselage connection, is slightly heavier than that of the Cessna singles. This results in less skin panel noise during turbulent conditions and, unfortunately, also in somewhat less useful load for the 150-hp Cherokee Cruiser compared with a Cessna 172 of similar power.

As a final comparison item, the Cherokees may tend to float a bit in ground effect and so are a bit more difficult to land than the high-wing Cessnas, particularly when approach speeds are a few miles per hour high.

Stepping back and weighing the various positive and negative features of each design, it is apparent that when given equivalent performance, powerplant reliability, maintenance requirements, and price, owners prefer an airplane offering ease of entry, slightly greater longitudinal stability, and downward visibility in comparison with one having superior visibility in the landing pattern but with somewhat less longitudinal stability, useful load, and ease of entry.

On balance there is really little difference between the Cherokee and Cessna singles; both configurations provide excellent aircraft meeting most of the design and maintenance criteria outlined. But another manufacturer also has a very popular single-engine model, although of a very different breed, the Beech Bonanza of Fig. 1-5. In production since 1947, this high-performance airplane is the acknowledged Rolls Royce of the general aviation market.

I remember attending the Cleveland Personal Aircraft Exhibition in 1946

FIGURE 1-5
Beech Bonanza V35B
Beech Aircraft Corp. Photo

with some Grumman associates and seeing the prototype Bonanza on display at a basic introductory price of $6900. I was then project engineer for a postwar personal aircraft design Roy Grumman believed would appeal to the many Navy pilots who had flown our rugged Grumman aircraft in the Pacific theater.

After seeing the Bonanza and comparing it with our similar design priced at $8100, we decided Grumman's military overhead precluded competition in the private aircraft market, and so our new design was promptly dropped. It was a nice airplane which performed well with standard controls and with the briefly popular two-control system (in which the ailerons and rudder were interconnected to a control wheel but the elevator control action remained separate). As things developed the Bonanza never sold for $6900, and for many years I have wondered if Grumman should not have entered the lightplane market back in 1946. Beech unknowingly effectively eliminated potential competition by posting an unrealistically low list price. The Bonanza remains in production today, but the equipped price may easily exceed *10* times the proposed 1946 list price—and Grumman has finally entered the general aviation market through the purchase of American Aviation.

In the years since first production back in 1947, Beech has produced more than 13,000 Bonanza aircraft of various model designations. This design is popular with the more experienced pilot because it combines speed with a proven structure and excellent handling characteristics. The Bonanza is a superb instrument airplane in the hands of a qualified pilot who flies regularly; as such this design fills a definite and specialized segment of the single-engine private airplane market.

Although requiring more pilot attention and maintenance than either of the leading volume Cessna and Piper aircraft because of the retractable landing gear and constant speed propeller, the Bonanza meets our basic design and most of the maintenance criteria. The owner pays a premium initially and for each flight hour in exchange for increased climb and cruising speed compared with the less sophisticated Cessna 172 and Cherokee aircraft. However, due to a combination of service reputation, declining dollar, and a well-organized owners association, the Bonanza enjoys an unusually high secondhand market value.

As we review these three leading types of personal aircraft, it is obvious that many factors in addition to detail design contribute to customer acceptance. Among these are effective sales, spare parts, and service organizations; the location of qualified maintenance and overhaul facilities in each sales area; a well-finished interior; an active secondhand market; and the known fact that pilots tend to purchase the type of airplane in which they learned to fly. Continued engineering development is also very important. For example, the 180-hp Mooney Ranger has been developed into the 200-hp, higher-performance Executive model of Fig. 1-6, offering a new level of performance to former Mooney customers and opening a new sales market for the company. Similar power and performance changes are also common to the Beech,

FIGURE 1-6
The 200-hp Mooney Executive
Mooney Aircraft Company Photo

Cessna, Grumman American, Piper, and Rockwell International families of private aircraft, providing expanded product lines and sales markets.

According to latest AOPA (Aircraft Owners and Pilots Association) tabulations, the average private pilot only flies about 45 hr per year. It is therefore essential that aircraft designed for the private flying market be easily handled in all flight conditions. To summarize our basic design considerations, and at the risk of seeming repetitive, our criteria require an acceptable airplane to be stable in flight at a given trim setting; exhibit no peculiar or demanding stall or ground handling characteristics; be easily controlled in the approach; and above all, exhibit *positive longitudinal stability* characteristics—as demonstrated by rapid damping of pitch oscillations in turbulent air and by *positive stick forces* (increasing control stick load with increasing pitch attitude). Of course, the structure must be well tested and of proven design, while the powerplant should be relatively simple to operate, free of restricted (red band) rpm settings on the tachometer dial, and not subject to induction system or carburetor icing. In other words, the private airplane must be simple to fly.

*   *   *

As we proceed through this book, references to other design material will be found in most chapters. Individual references are related to each chapter by notation such as 2.1, 2.2, 2.3; 3.1, etc. The first number refers to the chapter, and the one following the decimal point indicates the reference number for that chapter. As a further thought, please realize that all examples presented are intended to illustrate a point or problem and, as such, may not necessarily represent precise engineering solutions—particularly since we are holding all calculations to simple algebra or lower levels of mathematics.

Within these guidelines let us now move on to aircraft design, starting with the origin of structural loads, then reviewing the ever continuing search for utility as represented by special aircraft configurations, and finally getting to the root of airplane design through a study of desirable operational requirements and how they are realized. The balance of the text will show why various design options must be combined or traded off to satisfy these needs, while also meeting FAA design and type certification criteria. Our aircraft are designed, developed, and refined through just such a process, usually extending over a 2 to 3-year period.

2 THE ORIGIN OF LOADS

Private aircraft are certified in the United States under FAA Federal Air Regulations (FAR) Part 23 Airworthiness Standards (Ref. 2.1). These design requirements are accepted by practically every country in the world, although some foreign agencies (particularly in Australia and France) reserve the right to review design and flight data for imported aircraft before issuing an airworthiness approval. As part of a reciprocal agreement arranged at the Department of State level, the FAA accepts aircraft design standards of many other governments upon an "equivalent basis" for import of private aircraft into this country (Ref. 2.2).

FAA specifications have grown in volume over the years and now cover applied loads, airplane design, and flight test requirements in great detail. I first became acquainted with Air Commerce Manual 04, entitled *Airframe Airworthiness,* back in 1939. At that time ACM 04 cost 15 cents and totaled 140 pages that included the following: design terms and definitions; specific design requirements; static and flight test procedures; aerodynamic and applied loads calculations; weight and balance calculations; and aircraft modification and licensing procedures. The 1976 regulations cover the same ground with six different parts costing $18.30 and totaling well over 300 pages without including recommended procedures of any sort; in fact, Part 23 lightplane design requirements alone total more than 100 pages—so the quality and safety of small aircraft should have improved if the quantity of design specifications is any criterion.

Before proceeding further, a few simple definitions will be helpful in explaining design load terminology. The *load factor* applied to an airplane refers to the number of times the *applied loading* exceeds the 1 g weight. For example: a 10-lb radio subjected to a 4 g load factor applies a load of 40 lb to its supporting structure. So the *applied load* is 40 lb.

Now the load factor may also be referred to as an *applied load factor,* a *limit load factor,* or an *ultimate load factor.* The applied and limit load factors represent the actual design loads experienced during operation divided by the weight at 1 g; thus we obtain a load factor of 4 for our radio support example above. The ultimate load factor is obtained by multiplying the limit load factor by a 1.50 margin of safety. So the 4 g limit load factor would become a 6 g ultimate load factor, while the 40-lb limit load becomes a 60-lb ultimate load acting upon the support structure.

We also have *maneuvering* and *gust load factors* which may be *limit maneuvering, limit gust, ultimate maneuvering,* or *ultimate gust load factors.* In each case the 1.50 margin, known as the *factor of safety,* changes limit to ultimate, and the ultimate loads are used to design the aircraft structure—*creating a 50 percent overload safety margin.*

The meaning of the term *airframe* is another definition of interest to us at this point. The airframe is the basic aircraft structure that must be designed to support the applied loads; this includes the wings, tail surfaces, fuselage or hull, cowling, engine mount, control system, and landing gear. When all airframe components are assembled into one structure they support the powerplant, passengers, baggage, and fuel. So whenever the word *airframe* is used it refers to the load-carrying structure of the airplane, thereby excluding such items as the engine, propeller, powerplant controls, instruments, radios, hydraulic and electrical systems, and any special equipment such as oxygen or pressurization systems. This definition should be fully understood since the airframe is referred to quite frequently in technical aircraft books.

Because basic airframe weight varies directly with the design loads selected, our real interest lies in determining the maximum value of the *design limit load factor,* that is, the maximum actual loading the airplane will experience during any anticipated flight condition. This depends largely upon the purpose for which the airplane is being designed, with minimum *applied load factor* values established by FAA design requirements. For example, our pleasure and business aircraft are designed for normal category operation at 3.8 g limit load, while an acrobatic category airplane must be capable of stunt flying at 6 g limit. Utility category aircraft intended for rugged bush-type operation are stressed for 4.4 g limit load conditions, with high-performance sailplanes designed to 5.33 g limit load factors and restricted category aircraft meeting at least 4.2 g limit (Ref. 2.3).

The applied loads are in turn distributed into the skins, ribs, and beams or framing of the airplane structure through a rather complex process of detail calculations. So we see that the size and thickness of various wing, tail, fuselage, and landing gear members, including exterior skins of all-metal aircraft, are determined by the stress levels developed from Part 23 applied loads specifications.

Since the airplane is held together by bolts, rivets, and structural members designed by these applied loads, it is important that they are correctly determined for the intended use. Once flight limits have been established, the airplane can be placarded not to exceed certain design speed and weight limitations. But just what conditions create structural loads and how can operational limits be applied to aircraft design?

Aircraft loads originate during two distinctly different operating conditions, *flight* and *landing*. There is also a special combination of these loadings called the crash load condition which is now receiving deserved attention, as further discussed in Chap. 12. Since landing loads are covered for both land and water operation by Chaps. 14 and 15, we shall first examine the most important set of load conditions, the *flight* loads.

MANEUVERING LOADS   It so happens that flight loads are also of two types: *maneuvering* loads and *gust* loads. The word *maneuvering* does not necessarily imply aerobatic flight since such routine actions as a banked turn or a stall above landing speed are considered maneuvers in the sense that the airplane is subject to loads greater than 1 g. In level, trimmed, steady cruise flight all parts of the airplane and its contents are subject to a gravitational loading of 1 g. If you weigh 170 lb on your bathroom scale, you will also exert a measured loading of 170 lb upon your airplane in level flight. This loading is carried into the airframe through the seat and floor structure.

When you maneuver the airplane into a 2 g banked turn, your 170-lb body will load the seat and supporting structure at 340 lb instead of the original 170 lb, since 2 g × 170 = 340. In similar fashion the turning maneuver that doubles your body load also doubles the load applied to the wing and other parts of the airplane. Due to centrifugal action, greater lift is required to keep the airplane at a constant angle of bank and also at constant altitude, so a g loading develops. As shown by Fig. 2-1, for a banked turn the *limit maneuvering load factor* (which is the actual load on the airplane wings divided by the airplane weight at 1 g) increases with the angle of bank Θ in accordance with the relationship $F = W/\text{cosine } \Theta$. The force holding the airplane into a perfect turn is also a function of the bank angle and is given by the equation $L_H = F \sin \Theta$.

Note that if due to an uncoordinated turn the horizontal force L_H is greater than the centrifugal force, the airplane will move into the turn in a *slip;* if the horizontal component is less than the centrifugal force, the airplane moves out of the turn in a *skid* since there is not sufficient horizontal force L_H to satisfy balanced centrifugal load requirements. This is the analysis of how an intentional slip or skid can be developed.

Incidentally, for minimum drag in a turn all controls must be coordinated; as a visual presentation, the ball of the turn and bank (which is really the slip or skid) indicator will be centered. For a direct and cheaper approach to elimination of the slip/skid problem, a yaw string may be mounted near the base of the windshield directly ahead of the line of sight. Although some adjustment in string location may be necessary to allow for airflow around the windshield, this can be checked in trimmed, level flight to assure that the string aligns with the centerline of the airplane.

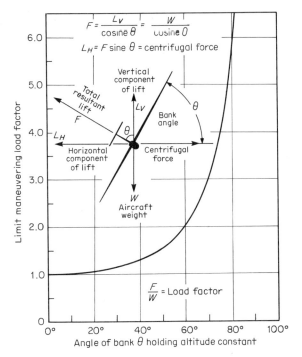

$$F = \frac{L_V}{\cosine\,\theta} = \frac{W}{\cosine\,0}$$

$L_H = F \sine\,\theta =$ centrifugal force

Vertical component of lift

Total resultant lift

L_V

Bank angle

F

θ

θ

L_H

Horizontal component of lift

Centrifugal force

W

Aircraft weight

$\frac{F}{W} =$ Load factor

Limit maneuvering load factor

6.0

5.0

4.0

3.0

2.0

1.0

0

0°   20°   40°   60°   80°   100°

Angle of bank θ holding altitude constant

FIGURE **2-1**
Variation of load factor with bank angle

In a slip or skid the yaw string will move *away* from the direction of the slip or skid; so "centering the string" will produce a coordinated maneuver. You may have noted that to minimize drag most sailplane pilots have a yaw string located directly in their line of vision, making yawed flight (a slip or skid condition) immediately apparent. Just to show that not everyone flies the same way, many sailplane pilots believe they experience less sink when carrying some slip in those steeply banked turns required for thermalling. Possibly this is so, due to some induced body lift.

The *limit maneuvering load factor* represents the actual load increase experienced by the airplane due to any flight maneuver. Basically this is represented by the increase in wing structure loading compared with steady 1 g flight. In our example of a 2 g turn the wing structure of a 2000-lb airplane would be loaded to 4000 lb. The limit load factor is then said to be 2. Referring to Fig. 2-1 again, it is seen that a 60° bank angle induces a 2 g limit load factor. Since the cosine of 60° is 0.50, by simple division $F = W/\cosine\,\theta$ becomes $F = W/0.50 = 2W$. Since twice the weight of the airplane is 2 g, we have shown that the 60° bank angle limit maneuvering load factor of Fig. 2-1 is correct.

As you realize from experience, flight loads can develop from many maneuvers in addition to turning flight. For example, elevator displacement creates a load on the horizontal tail which in turn loads the fuselage structure. Elevator action also changes the longitudinal trim, or pitch attitude, of the airplane; this action in turn changes the angle of attack of the wing and so varies the wing loading. Rudder deflection loads the fin and fuselage structure in side bending and torsion. A combination of aileron, elevator, and rudder action loads everything at the same time; so structural analysis of the airframe can become quite complex.

Aerobatics, of course, load the airplane from propeller to tail light, particularly if not properly performed. Biplanes are popular for aerobatic work because their short span reduces damping in roll which increases rate of roll for all maneuvers, while four wing panels offer the opportunity to mount four ailerons to further increase roll rate. And the biplane is also popular because the drag associated with brace wires and struts, combined with the lower structural weight possible from a biplane configuration, will prevent a rapid buildup in speed between maneuvers or during sloppily performed maneuvers.

Since maneuvering load factors vary with the square of the speed ratio because lift is a function of velocity squared (V^2), the inherent speed-limiting capability of the biplane is an essential safety ingredient for aerobatic flight. For example, if you stall at 60 mph in smooth air, the load factor will be 1 g—that is, when the airplane stops flying it has no external forces acting upon it and so will free-fall similar to a steel ball. At a speed of 120 mph, twice the basic stalling speed for this example, the load factor developed at stall will be 2^2 or 4 g. Applied to actual flying conditions this means: When caught in extremely turbulent air, simply reducing cruise a bit below twice the stall speed will prevent your airframe from being loaded at more than 4 g—regardless of the attitudes imposed upon the airplane. However, in extreme turbulence it is possible to experience a 4 g stall and spin-in due to maneuver loads arising from corrective control action, or to experience vertical gusts of such intensity that a wing panel can literally be sheared off. So regardless of speed/load considerations, avoid thunderstorms whenever possible; and if caught, reduce speed to twice stall and do not overcontrol to maintain attitude or altitude for the reasons just discussed.

In similar fashion, high-performance sailplanes can experience extreme maneuver loadings in the hands of an inexperienced pilot. In addition to indicating that a sailplane will go 40 ft along the glidepath for each foot of altitude lost, a 40 to 1 glide ratio also means that an 800-lb sailplane has only 20 lb of drag. It takes very little diversion of attention and less time for inexperienced pilots to find themselves moving at 120 mph out of 50 mph thermalling flight. If the sailplane stalls at 40 mph, any sudden pull back on the stick to reduce flight speed rapidly would load the wings to 9 g in this situation ($120/40 = 3$, and $3^2 = 9$ g). The result—bent wings at least, and possibly none at all.

The important things to remember about maneuvering loads are: (1) they arise from control action, (2) they increase with the square of the speed/stall ratio, and (3) they act upon everything in the airplane—airframe, passengers, engine installation, battery mount, baggage, etc.

To graphically establish previously mentioned structural design and flight placard speed limits the *V-n*, or really the *V-g*, diagram of Fig. 2-2 has been

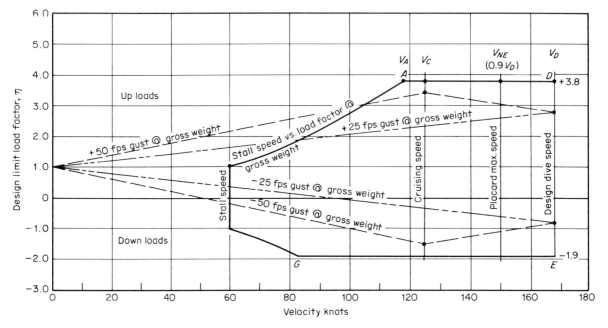

FIGURE 2-2
The V-n diagram for a normal
category medium perform-
ance airplane

developed. Similar diagrams are included in many airplane flight manuals. Since aircraft terminology refers to load factors as *n* values to prevent confusion with equations containing the gravitational constant *g* (of 32.2 ft/sec²), load factors throughout this book will be referred to as *n* values to agree with FAA and industry use. Also remember that in our review the term *limit load* is used to distinguish between the actual applied load required by regulation (or obtained by flight measurement) and the 50 percent greater *ultimate load* value used for detail design of each structural component.

One of the few design classifications permitting legal flight at an overweight condition is the restricted category, so designated by the FAA because operation of this class of airplane must be for specialized purposes and under a very restrictive license. When overloaded, such aircraft are using some of that 50 percent margin of safety designed into every airplane. As a result, restricted category aircraft are frequently denied the right to fly over populated areas or to carry passengers. Agricultural spraying and dusting aircraft are generally licensed in the restricted category under (Civil Aeronautics Manual) CAM 8 *Aircraft Airworthiness Standards,* and are permitted to fly at a weight above licensed gross weight in accordance with Fig. 2-3 (taken from CAM 8, Fig. 7-1).

It is not advisable to fly at more than design gross weight in private aircraft, and any mention of CAM 8 overweight provisions should not be construed to suggest such action. Agricultural (Ag) pilots and other restricted category aircraft pilots are usually experienced professionals who fly continually, have excellent control coordination, and risk no one's neck but their own. Whether designing or flying a Part 23 normal category airplane, remember that the 1.50

factor of safety has been included to cover special conditions such as locally marginal material properties and assembly, sudden maneuvers, mishandling in flight, severe turbulence, and the effects of old age—but not intentional overloading which also reduces climb performance and flight safety.

GUST LOADS   Despite the 1.50 structural safety margin, properly handled aircraft occasionally experience airframe failures from what are known as *gust loads,* our second flight load condition. Gust loads arise from the action of turbulent air upon the airplane and may be induced by vertical or horizontal air mass movement.

We are all familiar with thermals which produce vertical gusts of a generally minor nature, although they are representative of the type of flow producing major gust loads upon the airplane. Frontal action preceding a storm, vertical air motion resulting from winds blowing at right angles to a line of hills (ridge lift), clear air turbulence at altitude, and wind rotors associated with lenticular cloud formations are all capable of producing severe and sharp structural loads. FAA design requirements have recently been revised to consider more fully gust effects upon the airframe, and now require the structure to withstand 50 ft/sec gusts at cruise and 25 ft/sec gusts at maximum design speed.

When you consider that a 50 ft/sec vertical gust represents the impact of an air column rising at 3000 ft/min (34 mph), the energy contained in gust action becomes more apparent. Fortunately, private aircraft are generally designed with comparatively light wing loadings of about 14 lb/sq ft and so ride gusts in a manner similar to a cork on a wave crest. However, business and commercial jets experience sharp gust loads due to higher wing loading and greater airplane interia; so jet passengers are more comfortable, but the airframe is not.

Gusts may also act in a horizontal direction and can be particularly apparent and annoying during a landing in frontal passage conditions. *Wind shear,* a comparatively new term in our flight weather vocabulary, raises its head in these circumstances and is usually discounted by approaching at higher than normal speeds. Added speed means safety when the horizontal gusts range 20

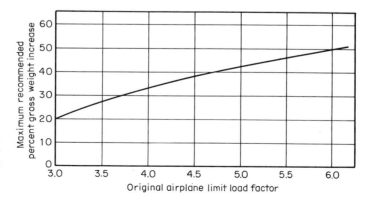

FIGURE 2-3
Restricted category permissible weight increase

to 30 mph above the average wind speed, since any sudden drop in gust velocity could stall an airplane approaching at normal speeds. For example, if stall occurs at 60 mph and the approach is at 80 with surface winds gusting to 45, a sudden 20 mph drop in headwind velocity back to 25 mph will cause the airplane to stall (80 mph indicated approach speed minus 20 mph drop in headwind velocity equal 60 mph stall speed).

This horizontal gust effect frequently occurs near the ground and can drastically affect approach glidepath angle as well as stall margin. If an approach is established along a 4° glidepath, a 20 mph headwind increase will steepen the angle to about 7°—requiring corrective action if the gust continues for any period of time. Under this set of conditions the tendency is to raise the nose, reducing airspeed and descent rate—but heading directly into a stall/spin condition if the gust suddenly terminates. Unfortunately, this situation is quite typical of the approach accident. For safe operation, plan to increase your normal approach speed during gust conditions by an amount equal to the reported gust intensity. For example, if runway winds are reported at 10 gusting to 20, increase your approach speed by 10 knots.

So we see that in addition to loading the aircraft structure, horizontal and vertical gusts also affect flight handling and safety. These conditions can exist at altitude as well as near the surface, although gust intensity is usually lower above 20,000 ft.

THE FLIGHT ENVELOPE

Now that maneuvering and gust loads have been described, let us return to the *V-n* diagram of Fig. 2-2. This is a typical flight envelope for a moderate cruising speed general aviation airplane. The maximum positive limit load factor is + 3.8 g running from the stalling speed at the load factor, Point A, to the *design dive speed* V_D of Point D.

The value of V_D is established by FAA requirements based upon the *design cruising speed* V_C. In accordance with FAR 23.335, V_C must be established as a function of the wing loading, unless a lower value of V_C can be substantiated by aerodynamic calculation or flight test demonstration. For normal category aircraft FAA criteria set $V_D = 1.40\ V_C$; for utility category operation, $V_D = 1.50\ V_C$; and acrobatic aircraft are designed with $V_D = 1.55\ V_C$. These are important basic design speed values since V_C and V_D will be used to obtain wing structure loadings, while the airplane must be demonstrated to be free from flutter at V_D.

The never exceed speed, which is the placard red line speed, V_{NE}, is established at 0.9 V_D by FAA design requirements. As a safety feature every production airplane must be factory dive-tested to V_D, which is a speed 10 percent greater than the owner may legally fly.

Gust load conditions are shown by the dash lines running to cruise speed V_C and then to V_D. While gust conditions are not critical for the wing structure of the airplane represented by Fig. 2-2 since their value is below + 3.8 load factor, gust loads probably will design the horizontal and vertical tail surfaces as they do for most modern aircraft. Tail loading will be reviewed at greater length in Chap. 7.

The negative loadings of Fig. 2-2 are half the positive values, indicating this diagram applies to a nonacrobatic, normal category airplane. This type of loading is particularly important to strut-braced high-wing monoplanes which have the strut placed in compression by any down load on the wings. If the down loads were as great as the up loads, the support strut could become of considerable size and weight. Fortunately, flight load investigations have indicated the range of $+3.8$ to -1.9 limit load factors to be safe for normal aircraft operation. However, for simplification of wing beam construction plus a bit of extra structural safety, cantilever monoplanes can be designed at slight additional weight to have equal strength under up or down wing load conditions; that is, wings capable of $+3.8$ or -3.8 limit load factors. This feature should be considered for amphibian aircraft equipped with wing floats, since floats transfer repeated water impact and structural rebound fatigue loads directly into the wing structure.

While load and speed values will vary from one airplane to another, the flight envelope diagram of Fig. 2-2 is representative for all modern aircraft and is the basis for airframe structural calculations, since it documents the design load factors at all permissible flight speeds and weights.

LEVEL FLIGHT LOADS

The airplane is loaded not only during maneuvering and gust conditions but in level flight as well. When all loads are in balance, level flight is maintained at constant airspeed and altitude. The forces acting in trimmed flight are shown in Fig. 2-4 for two different thrust line locations.

Most aviation texts refer to the four forces of flight: thrust, drag, lift, and weight. However there is a fifth force which interacts with the basic four; this is the horizontal tail balancing load represented by the symbol B of Fig. 2-4. This value must be carried by the wings when the tail load is down, since both weight W and tail load B must be balanced by wing lift L if the airplane is to maintain level flight at constant altitude.

Obviously, the lower the tail load the lower the total wing lift required; and less wing lift means less wing drag which results in more cruising speed. As a result, the search for reduced horizontal tail loads has produced lifting aft surfaces (the Flying Flea and similar tandem designs) and the canard configuration (which is equipped with a lifting tail located ahead of the wings as shown in Fig. 3-4).

Most modern aircraft develop relatively low horizontal tail loads by using main wing airfoils having low *section moment coefficient* values. The *section moment coefficient* $C_{M_{ac}}$ represents the summation of all aerodynamic forces acting upon the airfoil. These are mainly: (1) the components of section lift and drag pressures resulting from airfoil surface curvature and (2) the surface friction

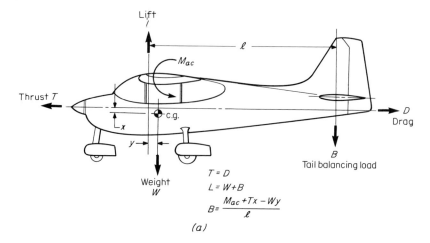

$$T = D$$
$$L = W + B$$
$$B = \frac{M_{ac} + Tx - Wy}{\ell}$$

(a)

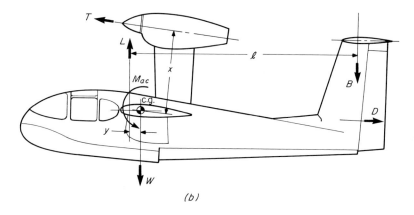

(b)

FIGURE 2-4
Level flight loads. (a) Forces acting upon the normal private airplane; trimmed cruising flight condition. (b) Forces acting with an offset thrust line.

drag unbalance between the upper and lower surfaces acting as an aerodynamic couple. This effect is represented by Fig. 2-5, which shows why the wing section moment M_{ac} normally acts to rotate a wing leading edge down (creating a counterclockwise moment).

The wing section moment acts about the *aerodynamic center* of the airfoil selected, the aerodynamic center being the locus of all external aerodynamic forces acting upon the wing section as shown in Fig. 2-5. The aerodynamic center differs slightly in location from the airfoil *center of pressure,* since the center of pressure is determined by summation of only the pressure forces acting upon the wing. As such the center of pressure is an important indicator of lift movement along the wing chord with variation in angle of attack and airspeed, while forces acting about the aerodynamic center create the torsional loads acting upon the wing structure. For most applications, both centers can be safely approximated at the same location of 25 percent wing chord aft of the leading edge.

M_{ac} is a function of section moment coefficient, air density, wing area, airspeed squared, and wing chord; it increases with any one of these factors as shown by the formula of Ref. 10.2. Because the wing section moment tries to rotate the wing leading edge down, the resulting twisting load acts upon the wing structure and the wing-to-fuselage connection to provide the torsional design loading for this important joint. In order to keep the fuselage as well as the wing from nosing down, a balancing force must be developed, such as the horizontal tail load B of Fig. 2-4. So the wing moment value M_{ac} has an important effect upon the entire airplane design.

Looking at Fig. 2-4b it is quite obvious that thrust line position with respect to the airplane center of gravity will also act to load or unload the horizontal tail. The T tail tends to be self-balancing for such thrust line locations, however, since tail-down load will increase with slipstream flow to counterbalance the nose-down moment resulting from power increases. Regardless of the degree of self-trimming obtained, the additional balancing tail load must be supported by the wings if level flight is to be maintained.

A specialized approach that reduces horizontal tail loads is used in sailplane design. Since power changes are not a problem, the normal center of gravity is moved as far aft as flight control requirements will permit—usually to the point of neutral stability where the airplane demands constant pilot attention for controlled flight. Looking at the diagrams and the B value formula of Fig. 2-4 we can see why the tail load decreases as the value of y increases; lift L acting upward ahead of the airplane center of gravity tends to overcome the nose-down wing section moment M_{ac}. For sailplanes, a combination of aft center of gravity plus a long tail arm (l value) results in low tail load and less wing support load, reducing required lift and wing drag values. This load arrangement results in low drag flight and reduced airframe loads, providing an efficient design configuration. Similar load optimization is necessary for all successful aircraft and

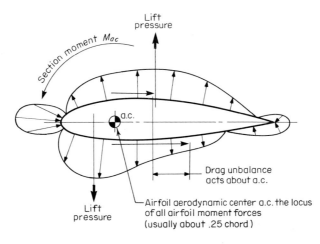

FIGURE 2-5
Airfoil section forces produce counterclockwise wing torsional moments.

requires considerable study during the preliminary design phase of every new aircraft program.

Before concentrating upon the design features, performance requirements, and handling characteristics peculiar to the single-engine private airplane, some additional design background will be most helpful. We shall first review the persistent design variations that never quite make the grade in an endless search for utility, and then move on to a brief discussion of successful configurations that are always popular.

REFERENCES 2.1 FAA Federal Air Regulations (FAR) Part 23; Airworthiness Standards: Normal, Utility, and Acrobatic Category Airplanes. Available from the

> Superintendent of Documents
> U.S. Government Printing Office
> Washington, D. C. 20402

Price $3.55 on a subscription basis providing all amendments as issued.

2.2 Appendix 2 of FAA Advisory Circular AC No. 21–2B: *Export Airworthiness Approval Procedures.*

2.3 FAR 23.337 for powered aircraft; the FAA *Basic Glider Criteria* handbook for sailplanes and gliders; and Civil Aeronautics Manual (CAM) 8 for restricted category aircraft.

3 THE SEARCH FOR UTILITY

With the exception of pilots primarily interested in flying as a relaxing or challenging sport, the private airplane basically appeals to those who can save time or expand their operating range through its use. Both reasons are really the same and relate to cruising speed plus the ability to fly more or less in a straight line between terminal points. For such users the private airplane provides a form of transportation not possible with any other vehicle. In addition, for a trip of any distance, the airplane offers less fatiguing travel. Unfortunately, the limit of aircraft utility is reached upon arrival at the destination airport.

Any further travel necessary to make a business call or visit friends will depend upon the availability of a taxi or rental car or being met by someone—all of which require some degree of prior coordination and some loss of time. In addition, weather frequently intercedes to terminate a flight before it begins or to cause en route delays.

In an effort to overcome these drawbacks inherent in present aircraft designs and to further increase private aircraft utility, the quest for a universal airplane capable of safe off-field landings, with the ability to travel highways, and with ever higher cruising speeds has appealed to all designers from the moment efficient horsepower first became available. However, the pathway to this goal is strewn with hundreds of failures of virtually all possible aircraft configurations. But along the way, a few types did evolve into today's accepted aircraft which will be reviewed in Chap. 4. Although we still lack the ultimate universal airplane, why have some designs survived while others have been rapidly rejected?

The answer apparently lies in some combination of the degree of operating skill required to pilot the airplane, the utility actually offered per dollar of initial cost, and the difficulty of obtaining FAA certification approval. Not surprisingly, a few basic design types seem to lead the list of repeated attempts to improve utility. By way of broadening our aviation background and with some slight hope of furthering progress, we shall briefly review the historically important, recurring "unconventional" configurations. Quite possibly a discussion of past problems will favorably influence future development.

REPEATED ATTEMPTS
AT PROGRESS
THE AUTOGYRO

The most persistent design is that of the autogyro. This type of rotating wing airplane was even popular back in the thirties when autogyro mail service was inaugurated in the Philadelphia area, then the center of rotor aircraft activity. The autogyro of Fig. 3-1 was developed into a roadable design that actually flew and was frequently driven on main highways during the mid-thirties.

Designed by the Autogyro Corporation of America (ACA), this small two-place ship was financed by the Bureau of Air Commerce for flight and road-handling evaluation. Although total production of all types was limited, autogyros were manufactured prior to World War II by Kellett Autogiro and Pennsylvania Aircraft Syndicate (Wilford Gyroplane) as well as ACA. This concept remains with us today in the two-place Umbaugh (later Air and Space Manufacturing) and J-2 (née Jovair, then McCulloch, and now Aero Resources) designs. As you can see, the recent models have had difficult production lives even though both received FAA Type Certificates under FAR Part 27. As of this writing, neither autogyro is in active production. However, the J-2 Gyroplane of Fig. 3-2 shows the modern trend in autogyro development.

Why does this design persist and why doesn't it find acceptance? The primary attraction is the ability to take off and land with full STOL (short takeoff and landing) performance. This capability may be expanded to achieve essentially vertical takeoff by driving the rotor to greater rpm than obtained in flight, then declutching and applying pitch to the rotor system. This sudden buildup in lift will literally snatch the autogyro about 50 ft into the air, at which time rapid application of forward power and the still lifting rotor will combine to provide forward flight. Of course, you might well ask why anyone would want the mechanical and control complexities necessary for jump takeoff of a design that

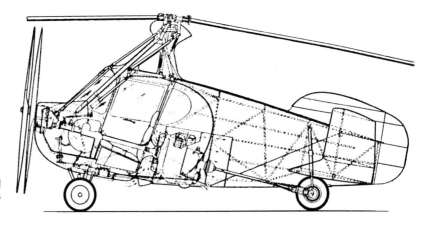

FIGURE 3-1
Roadable autogyro developed
during the thirties
Aviation Magazine

does not require rotor power for normal flight. This is a good question; and since not many people need such limited vertical flight capability, the related cost in weight and dollars tends to eliminate this feature of autogyro operation. For somewhat more initial cost, the owner requiring frequent vertical flight can purchase a small helicopter which will also have greater useful load.

In addition, the autogyro rotor system is more expensive to purchase and requires more maintenance than fixed wings. Even without a rotor drive system the controls are complex, while the combined weight of the rotor and pylon will exceed that of fixed wing area for an airplane of comparable gross weight. With somewhat different operational and handling techniques to be learned, thorough pilot checkout is mandatory for safe operation. As a final consideration, the autogyro will have a higher empty weight and lower useful load than a fixed wing airplane of comparable power. In fact the 180-hp J-2 useful load is only 510 lb.

Then what do we have left to benefit the autogyro owner? Really only the fact that the autogyro is capable of flying more slowly than a fixed wing airplane which permits a steeper *angle* of climb and lower landing speed. On balance the autogyro is a novelty approach that really does not offer the utility of a fixed wing landplane of comparable gross weight, power, or price. It has a natural market only among the few owners who must have some STOL capability at minimum price. As such, past autogyro designs really had nothing to offer and accordingly have never been successful—certified or not.

If someone should develop a simple rigid rotor autogyro having cruising speed, empty weight, and useful load similar to a fixed wing airplane of equivalent power, acceptance would change rapidly; but such a configuration appears fundamentally impossible.

ROADABLE AIRCRAFT
The attraction of being able to use the same vehicle to travel from home to a distant point is almost as persistent as the autogyro concept. Fostered by pioneer roadable autogyro development, the desire for door-to-door travel averaging more than 100 mph proliferated after World War II and remains alive today. Such companies as Convair, Fulton, Spratt, and Aerocar, plus numerous homebuilders, ventured into various types of roadable airplane development. Of these, only Aerocar ever received FAA certification, and that approval was for a limited production design. While Aerocar under the direction of Molt Taylor remains active in the development of roadable aircraft, the automobile is still the only road vehicle available. As would be expected, there are sound reasons why this is so.

In addition to the obvious economy of keeping your airplane at home, one of the great advantages offered by the roadable airplane is the ability to land ahead of approaching bad weather and continue traveling by road.

To be effective in combating weather in this manner, it is essential that all parts of the airplane continue the journey. Most of the past designers have hindered roadable development by overlooking this very fact. To leave wings, or wings and tail surfaces, behind during ground operation means that the journey will have to be retraced to reassemble the various flight components. This design approach is not practical simply because of the time lost in retrieval effort. In addition, if clear weather is reached by driving along the road, flight will not be possible with the wings left behind. And, unless housed, the temporarily discarded surfaces will be subject to wind damage and vandalism.

The FAA certified Aerocar of Fig. 3-3 has the optional capability of trailing all its components or leaving the flight structure in the garage or small hangar at the home airport. So the Aerocar can be an airplane, a roadable airplane, or simply a compact automobile.

Although most of the complex development required to reach this state of the roadable aircraft art is not apparent, the major problems are: steering on the road, which is much too sensitive with an aileron-type direct connection to the wheels; adequate road suspension plus retracting four wheels in flight and powering two for road operation; developing simply connected wing, tail, and propeller drive attachments; and cooling the buried engine. These features all add up to considerable vehicle weight and manufacturing cost. As a result, the roadable airplane offers superior, in fact, almost ultimate utility, but at in-

(a)

(b)

FIGURE 3-3
Aerocar III. (*a*) As ground ve-
hicle. (*b*) Ready for flight.
Aerocar, Inc. Photo

creased cost and decreased useful load compared to a conventional airplane of similar gross weight.

This is an appealing trade-off for many and one that would not preclude broad acceptance of the roadable airplane. Unfortunately, the real problems are more political than technical and begin with the transition from flight to ground operation. For at that point existing FAA flight standards along with federal and state motor vehicle regulations combine to defeat the roadable concept.

Because of weather conditions it may not be possible to land at an airport, presenting the likelihood that FAA airspace regulations will be violated if a road or roadside landing is made. But this prospect is minor compared with the situation faced upon reaching the highway. When operating as an automobile, the vehicle must comply fully with state motor vehicle laws which include: headlight, safety light, and exhaust emission standards; specified roadway design features; federal and state automobile taxes; liability insurance minimums; etc. To this may be added the present certainty of being arrested in most states for landing upon a public highway, to say nothing of probable local protest and active resentment. As Molt Taylor of Aerocar indicates, the political difficulties of roadable airplane operation in our changing society truly outweigh the technical development problems (Ref. 3.1).

In summary, although the design features associated with roadable aircraft development are formidable, they can be solved with existing technology to provide a useful vehicle. However, in the roadable airplane we find another instance where a combination of practical and legal complications have served to delay, if not prevent, introduction of an improved method of transportation. The FAA could materially assist in hastening roadable acceptance by acting as a focal point to define and develop roadable aircraft specifications and regulations that have the approval of air traffic control, federal highway development programs (roadside landing strips), state vehicle commissions, and various federal and state tax commissions. The task is difficult, but it must be accomplished before the roadable airplane can be economically produced. In view of this, it is unlikely that roadable aircraft can ever realize volume production—unless supported by the power of a major automotive corporation capable of amending present vehicle legislation.

CANARD AIRCRAFT The earliest powered aircraft had horizontal tail surfaces located ahead of the wings (Fig. 1-2). Practically all of the pioneers at one time in their early development employed this configuration, called *canard* by the French from some resemblance to a duck in flight. I do not know of any technical reason advanced for this initially universal tail arrangement used by such experimenters as the Wrights, Curtiss, Langley, Avro, Short, or the Farmans. The tail first arrangement seems particularly strange since Lilienthal, Chanute, Montgomery, and Pilcher, four technically sound basic experimenters, used an (now conventional) aft location for the horizontal and vertical tail surfaces of their 1890s gliders. Possibly the powered designers copied the Wrights' successful configuration, and then found this horizontal tail location a convenient reference for coordinating surface deflection with longitudinal trim response.

At any rate, the concept of a lifting forward tail is an appealing way to assist in carrying some of the airplane weight to reduce drag and increase cruising speed. Conventional airplane designs having tail surfaces located behind the wing use the horizontal tail to balance flight, thrust, and center-of-gravity travel loads. This usually means a down load on the tail as discussed in Chap. 2 and shown in Fig. 2-4, requiring an increase in lift coefficient to support the added wing load. Since wing drag increases with wing lift, we pay a climb and cruise penalty for the stability offered by an aft-located horizontal tail.

As shown by Fig. 3-4, a forward location of the horizontal tail divides the lift between two surfaces regardless of the load condition. If the main wing has flaps lowered, the added longitudinal moment is carried by added lift supplied by the forward (canard) wing. By increasing the *angle of incidence* of this section (the wing angle relative to the fuselage reference line), the needed trim lift is obtained without increasing the fuselage longitudinal trim angle. Of course, a greater trim angle increases the drag of the canard surface, but not to the degree resulting from an increase in fuselage trim angle. However, as a comparison with the balance formula for tail load B of Fig. 2-4 will indicate, determining the canard balancing load L_2 is a considerably more complex and critical problem than it is for the conventional aft tail location.

The canard arrangement offers the safety of a spinproof airplane if the canard surface can be designed to reach its maximum required trim position before stalling. When the wing incidence angle and center-of-gravity range are such that the main wing also will not stall at this canard setting, we have an airplane

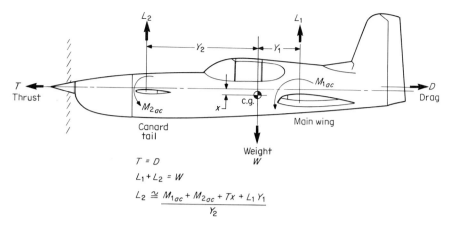

$$T = D$$
$$L_1 + L_2 = W$$
$$L_2 \cong \frac{M_{1ac} + M_{2ac} + Tx + L_1 Y_1}{Y_2}$$

FIGURE 3-4
Balancing forces acting upon a canard (horizontal tail forward) aircraft configuration

design that is both stall and spinproof. Unfortunately, this rather narrow design window is difficult to achieve. If the canard surface fully stalls in any type of flight maneuver, power "on" or power "off," the resulting loss of longitudinal stability can be disastrous. This is particularly true during climb-out or at approach speeds in turbulent air. The resulting tumbling may at best require considerable altitude for recovery, and at worst—and most likely—break up the airplane. The maneuvers accompanying canard wing stall can be similar to uncontrolled aerobatics.

The stability problems can be resolved, however, and canard interest has recently been rekindled by two new designs developed by Burt Rutan. The Vari-Viggan and the smaller VariEze two-place pusher canards have been specifically designed of composite wood, glass, and foam construction to permit ease of assembly by homebuilders. Shown in Fig. 3-5, the VariEze set a world's closed-course distance record at the Experimental Aircraft Association Annual Convention in Oshkosh, Wisconsin, during August 1975.

Possibly, this new airplane may indicate real progress for canard enthusiasts—at least in the field of very light aircraft. However, before sales volume sufficient to warrant certification can be anticipated, this design configuration must overcome opposition to reduced forward visibility, buyers' mistrust of an unusual or unconventional appearance, and the FAA flight stability requirements. For those seriously interested in canard design and stability criteria, an excellent basic paper is given in Ref. 3.2, while Rutan's work is covered by Ref. 3.3.

FLYING WINGS At one time new flying wing designs appeared in powered form on about a semiannual basis; their mission was to reduce drag and so increase cruising speed and operating efficiency. Waldo Waterman's work probably came the nearest to private aircraft certification, while Northrop Aircraft carried this con-

FIGURE 3-5
The canard VariEze in flight
*Rutan Aircraft Photo
by Don Dwiggens*

cept to the ultimate with the 172-ft span, 400-mph XB-35 built at the end of World War II. Repowered with jet engines, this ship became the YB-49 flying wing of 1950, but it never reached production.

The Vought XF5U-1 designed by Charles Zimmermann of NACA (National Advisory Committee for Aeronautics) was a particularly interesting version of the flying wing configuration in that the wing was essentially of double elliptical shape and of very low aspect ratio. I had the pleasure of working on the twin-propeller powerplant drive system for this design during 1942. Although having sufficient thrust to hover in flight (while looking like a pumpkin seed on edge), and successfully flown in prototype form, the XF5U-1 was abandoned when jet-powered fighters arrived with flight speeds beyond the capability of this maneuverable design.

The basic difficulty with the flying wing configuration is stability, really, the lack of stability about all three axes—pitch, roll, and yaw. Although very sophisticated autopilot controls were designed into the Northrop military wings, such devices exceed the cost and maintainability levels acceptable for private or even commercial aircraft.

We find the flying wing configuration kept alive today by private experimenters working with gliders, Rogallo wings, and small powered aircraft, although current FAA stability requirements preclude type certification of any simple design—and so this concept is not suitable for private aircraft production.

VTOL AIRCRAFT If a simple configuration could be developed, the VTOL (vertical takeoff and landing) airplane would be the ultimate flying machine. The ability to take off and land vertically with present aircraft-type control, powerplant, and propeller systems would provide maximum utility and safety. Imagine being able to come to a midair stop when encountering unexpected IFR (Instrument Flight Rules) conditions and then slowly descending to a convenient landing site until the weather improved.

This appeal is quite universal. Although the helicopter does all of this, it is expensive to purchase and to maintain, requires much greater pilot skill than fixed wing aircraft, and is not suitable for IFR operation unless very expensively equipped. As a result we have repeated attempts at some version of fixed wing VTOL flight.

The control and trim requirements during hover with rigid hub propellers are most severe, particularly in gusty wind conditions, so the control system tends to become complex. Quite possibly if we could be content to land in an area 100 yd square instead of 100 ft square the degree of control sophistication would be reduced. Again, stability raises its head as a barrier to acceptability. The stability problems associated with fixed wing VTOL operation exceed those of the flying wing, so some time will be required before such aircraft become available. But since this concept offers the most utility and safety of any airplane configuration, we might reasonably expect acceptable fixed wing VTOL aircraft to be developed within the next 25 years. Somewhere between now and the end of this century a marriage of the airplane and helicopter should occur, resulting in an easily handled rigid rotor design capable of vertical flight as well as zero forward speed at some useful altitude. This machine would combine practically all functions now expected of different light powered aircraft into one vehicle. Such a compound airplane could provide the next quantum jump in aeronautical development. Let us hope it will be produced at a price equivalent to modern private aircraft.

REFERENCES 3.1 M. B. Taylor, *The Aerocar Flying Automobile.* Society of Automotive Engineers Report 740393, April 1974.

3.2 Joseph V. Foa, "Proportioning a Canard Airplane for Longitudinal Stability and

Safety Against Stall," *Journal of the Aeronautical Sciences,* December 1942, 523–528.

3.3 Burt Rutan, Rutan Aircraft Factory, Bldg. 13, Mojave Airport, P. O. Box 656, Mojave, California 93501.

3.4 *Potential Structural Materials and Design Concepts for Light Aircraft.* NASA CR-1285, March 1969.

4 WHY THAT CONFIGURATION?

Like automobiles, aircraft designed for the same use are becoming so similar in appearance that it is extremely difficult for even an experienced observer to tell them apart. Of course, the high-wing designs can be separated from the low-wing models, but try to distinguish a Beech Musketeer from a Piper Cherokee in the landing pattern. And even all the high-wing Cessnas look much alike since they were derived from the same basic design.

While anyone interested in flying can readily tell a sailplane from a powered airplane, it practically takes an owner to separate and identify the various models of high-performance sailplanes. As another example of configuration similarity, the small biplane dominates the acrobatic field due to its ability to perform complex maneuvers. After some thought it becomes apparent that aircraft designed for similar use look alike because of the performance and structural demands peculiar to the service required.

As indicated by the title, the real purpose of this book lies in the analysis and description of how various design features and requirements affect aircraft performance and mission capability. Incidentally, for our purposes the terms *private* or *personal* aircraft refer to single-engine pleasure, business, and training planes, a type basic to the general aviation industry and familiar to every pilot.

In order to understand more fully the effect of design upon performance we shall briefly analyze the characteristics of major types of aircraft in current use. As private aircraft design features were covered in Chap. 1, we shall not review them again at this time. However we shall consider the different and special criteria for business twins, agricultural types, acrobatic and sport-flying aircraft, sailplanes, and seaplanes. Following this chapter we shall discuss in detail just how operational requirements actually mold the final aircraft design, presenting the various options in owner's manual or aircraft flight manual style. Just how and why design requirements result in specific airplane configurations tailored to the intended mission will become increasingly apparent as we review these different aircraft.

BUSINESS TWINS   The twin-engine, under 6000 lb gross weight airplane is one of the most useful types available for the general aviation market. Since such aircraft are primarily used to go someplace on schedule if legally possible, the design must perform acceptably under IFR conditions. Reliability and safety are important requirements, so two engines are quite logical, provided the airplane can maintain level flight or climb slightly on one engine. Icing protection is also necessary for practical operation.

Because two engines are used to assure flight safety in the event one fails, the critical design consideration will be ease of handling with one engine out. Stability under such adverse conditions becomes important, so a relatively long tail arm is necessary. A large fin is also required to control the yawing moment produced by the combination of offset thrust from the operating engine plus drag from the dead engine.

While mentioning engines, another important design requirement for the modern business twin is a fairly high cruising speed. To realize high cruise a low total power loading approaching 10 lb/hp plus a clean aerodynamic design will be necessary, and both are mandatory to obtain acceptable single-engine operation. Turbocharged engines make a somewhat more complicated powerplant installation, but they provide altitude power for improved cruise performance as well as a greater margin of power for single-engine operation at intermediate altitudes.

This type of airplane must have a range of 600 to 1000 statute mi in order to cover intermediate trip distances faster than commercial transports. Endurance is also necessary to meet alternative destination requirements when operating in actual IFR conditions; so considerable fuel capacity must be provided to feed both engines.

For passenger safety and reduction of wing bending loads, all fuel should be stored in the wings. Naturally, a mass of fuel distributed along the wing span and/or at the wing tips will act to reduce the rate of roll. As a result, the twin must have powerful ailerons to satisfy crosswind landing requirements when loaded, a consideration most recently demonstrated by Piper's increasing the aileron span for their 1976 model Seneca II of Fig. 4-1.

The requirement for high cruising speed will result in reducing wing area to lower total drag, which in turn produces higher landing and stall speeds—meaning powerful flaps are necessary. Although a comparatively high-wing loading will tend to reduce climb rate and ceiling, any such loss will be more than replaced by the increased climb power available from two engines plus low configuration drag.

Business twins are generally used as personal transport aircraft carrying four to six passengers, making adequate ventilation and superior soundproofing basic design considerations. Ground ventilation in particular can be critical, so it is not surprising that cabin air conditioning systems have been developed as optional equipment for modern twins. Due to IFR flight communication and navigation requirements, the instrument panel must have space for plenty of

FIGURE **4-1**
Piper Seneca II
Piper Aircraft Photo

electronic equipment and flight instruments. Of course, with present FAA airway and traffic control regulations the need for a great deal of electronics and instruments is becoming quite common for all aircraft and certainly is not limited to business twins.

Since this type of airplane is usually flown by experienced pilots who fly more frequently than the single-engine-rated private pilot, stall characteristics are not as critical as for private aircraft. In fact, a gentle or prolonged stall peak

FIGURE 4-2
Twin-engine Cessna 337
Skymaster
Cessna Aircraft Company Photo

will be of secondary importance to minimum section drag and moment characteristics, with modern designs using a flush-riveted laminar-flow airfoil wing.

Business twins generally operate from paved runways and at comparatively high takeoff and landing speeds; so retractable tricycle gear will be used for ease of ground handling, improved flight safety in crosswind conditions, and minimum cruise drag. However, for rough terrain and bush-country use, consideration should be given to tail-wheel-type landing gear, as will be discussed in Chaps. 6 and 14.

Although there are not too many practical locations possible for just one engine, twins do have two power packages that can be moved about. Cessna designers have developed a tandem-powered passenger pod locating the thrust of both engines on the aircraft centerline. As shown by Fig. 4-2, the Cessna 337 design will not experience yawed flight from offset thrust when either engine fails, although some flow change may require minor trim corrections. Since less yaw means less drag with the same power, this design can maintain a higher single-engine ceiling at gross weight than is possible with engines located off the centerline. Or conversely, under single-engine operation this configuration can carry greater weight to a given altitude than can an offset twin design of similar power.

Of course, designers have worked on other methods to reduce single-engine drag, not only to improve performance but also to reduce the cabin workload. One of the first and primary methods was the elimination of windmilling prop drag by *feathering* a dead engine propeller (so the blade section parallels the airflow). Opposite rotating propellers are sometimes used to balance airflow over the wing and tail, reducing trim changes with power variations for both twin-engine and one-engine-out conditions; but the feathering propeller remains the most productive aid currently available for single-engine operation.

With the foregoing design criteria, successful business twins rather expectedly have the following common features: high cruising speed and twin-engine range; IFR capability with icing protection; excellent stability plus comparative ease of handling with one engine stopped; relatively high power for climb and cruise performance as well as single-engine operation; long tail arm and large vertical tail for directional stability and single-engine control; low-wing area compared with private single-engine aircraft plus a low-drag airfoil section; long-range-cruise fuel capacity with fuel preferably located in the wings; powerful ailerons and flaps; feathering propellers, possibly counterrotating; tricycle-type landing gear; and superior cabin soundproofing with ground ventilation.

Although all production twins seem to follow this design format, I believe a market exists for owners who want two engines primarily for cross-country and IFR safety, rather than for great speed or altitude performance. Since they would only occasionally carry four people, this design could have a Cherokee 2 + 2 type cabin, two small 100- to 125-hp engines, constant speed propellers, fixed faired tricycle landing gear, and about a 600-mi maximum range. Feathered props should not be necessary since single-engine yaw forces would be low; elimination of feathering along with the fixed gear will reduce purchase and maintenance costs as well as cockpit workload. What more could the occasional pilot ask for?

Such a design would carry two plus much baggage and fuel at 150 mph in great safety over rough mountains and through nonicing IFR conditions, day and night—and at half the cost of present twin-engine designs. Quite possibly the present twin-engine design configuration needs more thought and some revision.

AGRICULTURAL AIRCRAFT   We now turn to a very specialized type of airplane design; actually, a valuable farm tool and one capable of repaying its purchase cost within 1 to 2 years—the agricultural (Ag) airplane. Since the prime function of this type is to spray, dust, or seed from very low altitude and over practically any type of surface, it is essential that the airplane be extemely maneuverable and have an instantly responsive engine. Maneuverability is mandatory for turning in short fields and around or over obstacles, while instantaneous power response is necessary for sharp pull-ups around these same obstructions or to clear rising terrain.

In the hands of an experienced pilot, the less stable an airplane the more maneuverable it will be—at least up to the point of *neutral stability*. Neutral stability occurs when the airplane continues flying stick free in whatever attitude it has been placed; the stick forces per g are very low, and there may be no increase in stick load with elevator displacement or increase in rudder pedal load with rudder deflection. In other words, the airplane does not tend to return to its original pitch, roll, or yaw attitude when it has been displaced.

As a result, aircraft that have *neutral* or slight *positive stability* do not fight the pilot to return to their original flight path, so they are easily and rapidly controlled in pitch and yaw—but they must be constantly flown. As a matter of interest, neutral or slight longitudinal stability is usually associated with a fairly short tail arm. Due to the high degree of maneuverability required, slightly positive (marginal) stability is a basic design parameter for successful agricultural aircraft and is one reason this type should only be operated by trained professional pilots.

Roll rate must also be high for the maneuverability required. Biplane wings are one solution offering plenty of area in a small span plus the opportunity to mount four ailerons. The short span reduces roll damping while two sets of ailerons increase roll power. Ag planes intended for small, enclosed-area operations such as rice-field control will be biplanes because of their maneuverability. One of the best of these, the Grumman American Ag Cat of Fig. 4-3, has been in production since 1958 with over 1800 aircraft delivered to date. Note that this design is quite short-coupled (short tail arm) and has plenty of wing area. In addition to providing a high rate of roll as well as great amounts of wing area, a biplane wing structure is lighter than a cantilever (unsupported) monoplane wing of similar area. Structural weight saved by the biplane configuration will increase payload and climb, or can be used to provide a more rugged and crash-resistant fuselage structure.

The Piper PA-36 Pawnee Brave of Fig. 4-4 will be noted to have a cantilever monoplane wing representing another design approach, since this Ag plane configuration with large wingspan is primarily intended to lay a wide swath across long and broad fields—which will not require tight maneuvering. Cessna and Rockwell also build low-wing monoplanes for the expanding agri-

FIGURE 4-3
Grumman American Ag Cat models G164A (left) and G164B
Grumman American Aviation Photo

FIGURE 4-4
Piper PA-36 Pawnee Brave
Piper Aircraft Photo

FIGURE 4-5
Grumman American Ag Cat
model G164B ready for
inspection or wash down
Grumman American Aviation Photo

cultural aircraft market, and many operators keep overhauling their old Stearman trainer biplanes which were the original post-World War II Ag conversions. So both monoplanes and biplanes are used, depending upon operational requirements.

Whether old or new, one wing or two, all agricultural aircraft fuselage structures have four things in common: the payload hopper is located ahead of the pilot so that it will absorb energy without crushing him or her in a crash; the structure is of welded tubing, again for crash protection; exterior panels are readily removable for maintenance and wash down of spray and dust chemicals (Fig. 4-5); and all tubular structures as well as control system components are thoroughly corrosion-proofed for protection from the chemicals being applied. In fact, the fuselage structures are designed and built so ruggedly that there are known cases of midair collisions from which both pilots were able to transfer to different airplanes and continue spraying, having completely demolished the first pair of aircraft.

Landing gear components must be equally rugged to withstand overload operation from rough fields. Since it does not take long to load and spread chemicals, the Ag plane operates on a short cycle, averaging more landings per hour than most aircraft with the possible exception of flight trainers. Conventional landing gear with the main wheels forward and a tail wheel is preferred for rough field work and is acceptable for professionally flown large aircraft.

The major source of Ag plane engines has been the World War II surplus market, with radials preferred since they are generally considered more throttle responsive than flat, horizontally opposed types. However, the required horsepower Pratt and Whitney, Wright Aero, Continental, and Jacobs radials are becoming scarce and therefore expensive. As a result other engine types are now being placed into production. The Piper PA-36 Pawnee Brave of Fig. 4-4 used the new Continental Tiara gear reduction flat six, while higher-powered turboprop engines such as the PT6A-34 are currently undergoing experimental flight evaluation on other aircraft.

Cruising speed is not important, since the Ag plane is not intended for cross-country travel and 90 mph is an acceptable speed for spraying or dusting. In similar manner climb is only important in regard to getting the maximum load off the ground, and altitude operation is not normally a design consideration. To carry, or rather, to stagger off with the huge payload of chemicals or fertilizers to be applied requires much horsepower and wing area, particularly on a hot, Gulf Coast summer day. So load-lifting capability is more important than low drag. Of course, dust and spray systems are an integral part and purpose of every Ag plane and differ slightly for each type in order to provide the optimum dispersal pattern. However, we will leave the subject of chemical distribution to specialists in that field.

The important criteria for agricultural aircraft designed for working tight areas are: slight positive stability plus high rate of roll for maneuverability; payload capability; operation from rough fields; a rugged airframe to reduce service and maintenance; accessibility for inspection and cleaning; and corrosion protection. These requirements are satisfied respectively by the following: a short tail arm and the biplane configuration of short span plus four ailerons; wing area and plenty of power to carry the payload; rugged conventional landing gear; welded steel tube fuselage; removable fuselage panels and wing inspection openings; and thoroughly primed and painted components, especially those subject to chemical spray or dust impingement.

Requirements are similar for the monoplane Ag plane designed to work the more open areas, except that maneuverability is not as important. As a result the tail arm will be longer, with adequate roll provided by a single set of large ailerons. Here we can clearly see how a variation in operational criteria has resulted in a configuration change between similar types of aircraft; the monoplane design was selected when operational requirements permitted because it is cheaper to build than a biplane.

ACROBATIC AND SPORT AIRCRAFT

The original pioneering spirit so essential to the early growth of aviation remains alive today in the sport-flying field. Pilots who enjoy flying for pleasure and relaxation are turning in increasing numbers to homebuilt and custom aircraft of every configuration—with the open cockpit acrobatic design rapidly becoming a status symbol for this group. And if you've ever flown cross-country in an open cockpit during late spring, you know why this type of unrestrained flight is so appealing and satisfying. However, just what criteria do we apply to define a sport airplane?

There are really two categories of aircraft in this market: the highly maneuverable acrobatic type and the stable, easily handled sports pilot's airplane. Although each satisfies large segments of the sport-flying market, their design and performance requirements are really at opposite ends of the scale.

Considering acrobatic airplane design, we should have neutral longitudinal and directional stability if the aircraft is to be highly maneuverable, so a short tail arm would be expected. Since the biplane offers wing area plus a high rate of roll for reasons discussed under Ag aircraft, this configuration is normally used; while well-balanced, responsive control surfaces are essential for smooth maneuvers.

In order to climb rapidly, light weight, plenty of power, and ample wing area are required. With low airframe weight and high power, balancing any airplane becomes a problem because of engine weight; so not surprisingly acrobatic aircraft have their engines jammed back as close to the center of gravity and the pilot as possible. As high speed is not essential, the drag of biplane wires and struts can be tolerated—and is an asset when making rapid transitional descents because the airplane will come down fast without picking up too much speed. To handle complex maneuvers which can load the entire structure, a limit load factor of 6 must be used to design the airframe.

For climbing with much power at low speed a large-diameter low-pitch or constant-speed prop is required, making acrobatic aircraft usually seem long legged. Because changes in flow direction felt by the pilot can be rapidly translated into maneuver corrections, and for the pure sport of it all, the open cockpit arrangement is preferred. To save weight, and again for the sport of it, the landing gear will normally be of conventional type with a tail wheel. The Pitts Special of Fig. 4-6 is an outstanding example of the modern acrobatic biplane, and one which has received many first-place aerobatic awards in the United States and Europe.

Of course, acrobatics can be taught in aircraft intended for other purposes, provided the structure has been properly designed for such use or the maximum weight has been sufficiently reduced to permit higher load factors. For example, an airplane designed to 3.8 g loading at 2300 lb (for normal category use) could be loaded to 3.8 $g \times 2300/6$ g = 1457 lb at 6 g. Although this weight might permit carrying little more than a student pilot plus instructor, two parachutes, and partial fuel, limited or full acrobatics would be possible

FIGURE 4-6
Pitts Special acrobatic biplane
Pitts Aviation Enterprises Photo

provided the wing structure is capable of taking sufficient negative (down) loading. Acrobatics should never be performed in a normal category airplane unless the flight manual permits such operation at reduced weight—the reason being that if the structure has not been suitably reinforced, some unexpected and spectacular acrobatics will result if the wing fails under down loading.

A few standard production aircraft have been redesigned and strengthened for student acrobatic instruction, notably the Cessna 150 Aerobat and the Bellanca Citabria. Also, the Pitts biplane has been expanded to permit tandem seating for acrobatic instruction and pleasure flying. These three models are the current standard acrobatic training aircraft.

A new series of acrobatic aircraft is now emerging in a monoplane configuration. Although this type has been more popular in Eastern Europe than in the United States, development models of acrobatic monoplanes are being built and flown here with some success. However, it would seem doubtful that any monoplane could offer maneuverability superior to that of a biplane in the hands of a skilled acrobatic pilot.

Based upon competition results, the highly successful acrobatic airplane has small span biplane wings for maximum area and rate of roll, light structural weight, and a short coupled fuselage mounting a large engine close to the center of gravity. Because of the large-diameter propeller required to absorb power and provide a high rate of climb at low speed, the conventional landing gear tends to pitch the nose up at rest, which is also the three-point stall landing attitude. Include coordinated and responsive controls, an open cockpit for wind-flow sensing and pleasure, and we have the design formula for practically every acrobatic biplane design.

This concept has also been carried over into the pure sports pilot's airplane as indicated by the great amount of homebuilt biplane activity among Experimental Aircraft Association (EAA) members. Totaling over 40,000 active aircraft enthusiasts, the EAA has sponsored the design and sale of plans for constructing the easily handled ACRO Sport biplane and the Pober Pixie monoplane designed by EAA President Paul Poberezny shown in Fig. 4-7. Due to a combination of factory-built airplane cost, creative interest, and love of flying, there are literally thousands of homebuilt aircraft under construction at this time. Many designers offer plans and kits for a variety of biplane and monoplane aircraft of conventional and unconventional configuration.

To be safe and successful, the sports pilot's homebuilt airplane differs from the purely acrobatic type in that it should be easy to fly. Many homebuilders do not have any pilot rating and expect to learn to fly in the airplane they are carefully constructing; while others are only occasional pilots looking forward to the satisfaction of flying the airplane they are building. One of the best engineered and most efficient homebuilt landplane designs is Ladislao Pazmany's PL-2 shown in Fig. 4-8. This airplane has also been built in quantity as a trainer for the air forces of Taiwan, South Vietnam, South Korea, Indonesia, and Ceylon.

FIGURE 4-7
Experimental Aircraft Association Pober Pixie sportplane
Sport Aviation Association Photo by Lee Fray

FIGURE 4-8
Pazmany PL-2 homebuilt design
Pazmany Aircraft Photo

While this design is of all-metal construction, other types available are fabricated from wood, steel, or aluminum tubing with fabric or metal covering, plastic foam with fiber glass skin, and combinations of all methods. Through such experimental structural development the EAA membership is making a direct contribution to the private aircraft field, since some acceptable form of composite structure will eventually be evolved to reduce airframe construction labor-hours and cost. At the present time, FAA design criteria unfortunately tend to preclude or hinder the use of composite plastic construction for certificated production aircraft, but this situation could be changed if a demonstrably acceptable form of composite construction were developed.

An increasingly large and interesting part of sport flying centers around the reconstruction or duplication of "golden oldies"—antique aircraft produced during the late twenties or early thirties. While an unbelievable amount of time, energy, and money can be spent tracking down and rebuilding old components, the final results can be worth the effort as shown by Ed Wegner's Grand Champion Antique Award-winning American Eagle of Fig. 4-9.

Similar effort is expended to revive original Wacos, Great Lakes Trainers, DeHavilland Moths, Fairchild, Stinson, Taylorcraft, Piper, and Champion aircraft because all these planes are stable and relatively easy to fly. And that is the prime criterion for the sports pilot's airplane—ease of handling. This means adequate stability, gentle stall, low landing speed for short-field performance and good angle of climb, moderate cruise, and responsive controls.

To provide these characteristics the acceptable sport airplane will have a fairly low-wing loading varying from 6 to 12 lb/sq ft, a power loading of 10 to 22 lb/hp, combined wing and power loading totaling less than 30 and preferably under 25, long tail arm, positive longitudinal and directional stability with forces about equal for all control surfaces, and responsive controls. Useful load capability and cruise performance must be adequate, but are not of prime importance since this type of airplane will be used principally for local pleasure flying or short cross-country hops.

SAILPLANES Although this book is primarily concerned with airplanes, which by FAA definition are engine-driven fixed wing vehicles (Ref. 4.1), we should also include
some review of sailplane design since motorless aircraft fascinate everyone
—pilots and nonflying public alike.

Unpowered aircraft were originally called gliders because that was all they
could do—glide downhill. Due to a combination of high drag, inefficient wing
sections, relatively heavy wing loadings, and poor pilot technique the first gliders had very limited performance.

Following World War I, German aeronautical engineers were restricted to
motorless flight. In order to develop pilots for Goering's Luftwaffe, German
technical effort was concentrated upon the development and construction of
great numbers of increasingly sophisticated glider designs. Eventually the
state-of-the-art reached a level permitting sustained flight by use of ascending
air currents—and the sailplane was born.

Today's sailplanes fall into three categories: trainer, intermediate, and
high-performance or competition designs. Under proper conditions any of
these types is capable of soaring flight; they are principally defined by glide
ratio and the degree of pilot skill required for safe operation.

The Schweizer 2-33A of Fig. 4-10 is the standard United States two-place
trainer and demonstration ride sailplane. With steel tube, fabric-covered fuselage and tail surfaces plus an all-metal wing, the 2-33A is sufficiently light to
soar readily with one aboard, and has taken two people more than 10,000 ft
above the launch point in strong lenticular wave lift conditions. This sailplane
has a glide ratio approaching 20 to 1 at 45 mph and is relatively easy to fly. The
stall speed of about 35 mph indicates that things happen slowly and with minimum impact energy. As a result, a pilot has to be quite lax to get into serious
difficulty with this trainer.

The intermediate class sailplane is usually a single-place design offering a
glide ratio between 25 to 1 and 30 to 1. This type is capable of limited

FIGURE **4-10**
Schweizer 2-33A training
sailplane
Schweizer Aircraft Photo

FIGURE **4-11**
Schweizer 1-35 high-
performance sailplane
Schweizer Aircraft Photo

cross-country flight as well as ridge and thermal soaring and may be equipped with oxygen for altitude work. The Schweizer 1-26 sailplane is representative of this category and constitutes the largest single class of active one-design sailplanes in the United States. In fact, 1-26 regional meets and national contests are held in very similar fashion to one-design yacht competitions. The intermediate sailplane is somewhat more responsive and sensitive than a trainer and will require a higher level of in-flight planning.

The high-performance design is usually a single-place type, as represented by Fig. 4-11, and will have a glide ratio between 30 to 1 and 40 to 1. Along with this high glide ratio we find that things happen very fast and also that elevator control response is quite rapid and sensitive. In order to obtain minimum drag the tail load has been kept very low, resulting in essentially neutral or slightly positive longitudinal stability. As mentioned in Chap. 2, at a 40 to 1 glide ratio an 800-lb sailplane has only 20 lb of drag. So it is slippery and capable of rapidly picking up speed, which can be quickly converted into high g loadings in the hands of a novice sailplane pilot.

Regardless of the design category, certain features are common to all sailplanes. Since it is necessary to remain in a rising thermal as long as possible to gain altitude, and as most thermals are of small diameter, the ability to circle tightly and slowly with minimum drag is most important. Higher-performance designs achieve this by use of high aspect ratio wings, long tail arms, and neutral stability. The resulting light elevator forces are also of benefit as they are less tiring during long duration flights. However, neutral stability flight requires a relatively high degree of pilot skill, so increased longitudinal stability with its associated higher drag would be expected for trainer and intermediate designs compared with competition sailplanes.

High aspect ratio is also an important factor in obtaining high lift to drag ratio flight, because drag caused by wing lift is reduced in direct proportion to increased *aspect ratio* [which is the ratio of (wingspan)² divided by wing area]. This of course explains why sailplanes have large span, narrow chord wings; the designer is reducing drag by increasing wingspan for a given area. This subject is thoroughly covered in Chap. 6 and so will not receive further comment at this time.

Due to long span wings, *adverse yaw,* the tendency for aileron drag to turn the ship opposite to the desired direction, becomes an almost incurable problem; this condition is overcome by leading with the rudder and applying rudder as necessary during turning flight. To obtain maximum response at minimum drag penalty, a long tail arm and effective rudder are necessary. It would also seem that some additional design effort could be directed toward either reducing aileron adverse yaw or developing a more effective lateral-control system.

Because of relatively low sailplane drag and the accompanying ability to exceed placard maximum speed, some form of speed control is desirable. Such control has typically been included in the form of wing-mounted dive brakes capable of limiting speed in a dive condition. Adequate speed control has most recently been provided by plain trailing edge flaps that may be lowered slightly for thermalling, raised a bit to reduce wing section moment and drag during cruise, and dropped up to 90° for landing approach control.

Competition sailplanes also feature a retractable landing wheel, faired cockpit canopy, and water ballast to increase wing loading and cruising speed for a given *L/D* (lift-to-drag) value. Since these are cross-country aircraft, they will carry some radio and duplicate sensitive rate of climb (variometer) equipment.

Sailplanes have been developed to a point such that all designs now look very much alike. The successful high-performance single-place sailplane typically has an aspect ratio of over 20 to reduce wing drag, a tail arm of around 5 times the average wing chord, tail surfaces having a smaller percentage of wing area than is normal for powered aircraft due to the longer tail arm (basically reducing surface drag), neutral to slightly positive longitudinal stability at an aft center of gravity to reduce tail load and drag, responsive ailerons with a high rate of roll to permit thermalling flight, and some form of speed and approach (sink rate) control surfaces—either dive brakes or flaps. Of course, the fuselage and all intersections must be of excellent aerodynamic design for competition types, although this is a bit less critical for intermediate designs and much less so for trainers. With such rigid criteria it is no wonder that sailplanes look alike.

SEAPLANES   While I could go on for hours reviewing seaplane design criteria since amphibians have been my particular interest for 30 years, this type will be covered in depth in Chap. 15. For our immediate purposes, and to provide background for those readers not seriously interested in the design of watercraft, the three basic types of seaplanes will be discussed with a minimum of detail. These aircraft are the floatplane, the hull-type flying boat, and the amphibian, which may be of either float or hull configuration equipped with landing gear.

The Cessna 172 Skyhawk of Fig. 4-12 is a typical floatplane conversion of an

FIGURE 4-12
Cessna 172 Skyhawk with
EDO 89-2000 floats
EDO Commercial Corporation Photo

airplane originally designed for land operation. This conversion capability explains much of the popularity of float seaplanes; for relatively small additional cost an owner can enjoy water operation during the summer, returning to wheels or skis when freezing weather sets in. Many seaplane pilots consider the floatplane easier to dock than hull-type flying boats, further adding to acceptance of the floatplane. However, this is not necessarily so, as will be discussed in Chap. 15.

A hull flying boat would look like the Teal amphibian of Fig. 4-13 but with the landing gear removed. We cannot include a picture of a modern flying boat, since with the exception of the large ocean-crossing clipper types of the thirties and forties there really has been no pure hull seaplane produced since the late twenties. The small boat went out of production primarily because the cost of building a seaplane hull could not be justified for limited water operation. This consideration is amplified by the belief that landplanes can be mounted on floats to do the same job. But they cannot do the same job as a hull seaplane when operating in rough sea or severe wind conditions simply because the floatplane is poised high above the water with no outboard wing float protection—and therefore is much more vulnerable to capsizing than is the hull-type seaplane.

The hull seaplane survives today in the amphibian configuration capable of operating off land or water. This type of utility airplane was popular before and just after World War II in various twin-engine designs developed by Grumman, and remains in production today in single-engine versions such as the Skimmer, Buccaneer, and Teal amphibians produced by different small companies.

Basically a design concept offering a unique degree of utility through both land and water operation, the amphibian as shown in Figs. 4-13, 4-14, and 4-15 has a hull plus retractable wheels or wheels in floats for water operation.

FIGURE **4-13**
Teal amphibian with the first
T tail approved by FAA for
Part 23 aircraft
Howard Levy Photo

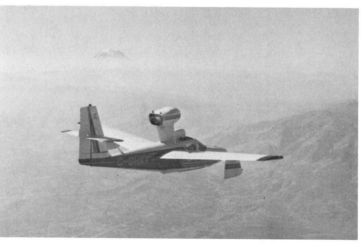

FIGURE **4-14**
The Lake Buccaneer am-
phibian
Lake Aircraft Photo

FIGURE **4-15**
Cessna 185 Skywagon with
EDO 597-2790 amphibian
floats
EDO Commercial Corporation Photo

Due to the hull chines and transverse step presently required for water work as illustrated on Fig. 15-1, amphibian drag is greater than that of a landplane of similar gross weight; and so the rate of climb and cruising speed are lower. Since water contact at 60 mph has an impact effect like that of a concrete surface, the hull bottom skin and frames must be much heavier than for a landplane structure, adding to hull weight and cost.

Floats with retractable wheels as shown in Fig. 4-15 are capable of providing amphibian operation through a fairly simple exchange of landing gear. However, amphibian floats are heavy and costly, generally resulting in a conversion that is more expensive than an amphibian flying boat of similar capacity—and much more difficult to handle in land crosswind landings or severe wind conditions on water because it sits so high off the surface.

Whether of float or hull type, seaplanes must be designed to function well on water since that is the primary reason for their existence. This means that modern seaplanes should have very mild or no *porpoising limit* characteristics (that is, no severe oscillation in pitch at high or low hull trim angles on the water during takeoff or landing); have no tendency to *skip* (pitch out of the water during landing); have the ability to perform step turns of small radius in order to operate from small bodies of water; be controllable on the water for precise sailing or steering when docking; and be able to take off with minimum water run on a hot day.

Adequate control at low speed is also essential during the unstable transition from a displacement boat to a planing seaplane. Since this takeoff phase, known as "getting onto the step," occurs around 20 mph, ailerons and elevator surfaces must be fairly large to be effective at this low speed.

All of this, naturally, costs in some fashion. Water performance requires power and wing area—there are no substitutes—and both are heavy and expensive. A watertight and rugged hull is also heavy and expensive, but landing on water at 60 mph is quite similar to landing on a runway with the gear retracted. So the hull must be strong. For operation around saltwater a corrosion-proofed structure and control system are essential, although more costly than similar landplane components. And the large movable surfaces required for low speed control may feel heavy at cruising speed.

Seaplanes are used principally by private or sports pilots and bush operators. This means a stable design is required, generally following the criteria set down for private aircraft. In addition, whether a hull flying boat or floatplane design, the seaplane will be more expensive and carry less useful load than a landplane of equivalent weight. To be effective, it should also have more horsepower. In return, the seaplane offers increased utility plus the joy and freedom of water flying. Possibly most important of all, the boat-type amphibian offers the combined safety of multiple landing fields and the crash protection of a rugged hull.

Despite inherent cost, cruising speed, and reduced useful load limitations, small amphibian performance has now been developed to a degree permitting limited volume production with three designs currently certified to FAA requirements. After many years of frequently discouraging effort, amphibian utility is slowly but steadily being recognized as sales increase from year to year.

Just as riverboats opened nineteenth-century frontiers, the amphibian will soon be used to expand commerce in the developing countries of Africa, Asia, and South America where lakes and inland waterways are plentiful but airports are not.

OPERATIONAL REQUIREMENTS

As this review should clearly indicate, a number of different aircraft designs are required to fill various operational missions. But whether aircraft are designed for private flying or specialized use, they must satisfy certain basic flight parameters if they are to receive pilot acceptance.

While the order of importance or weighting of items may vary with owner preference or actual use, the primary criteria used by the average general aviation pilot to evaluate powered aircraft are:

1. Crosswind handling

2. Rate of climb

3. Adequate stability for IFR operation and approach procedures

4. Useful load

5. Cruise performance

For some applications cruise performance may be more important than useful load, in which case criteria 4 and 5 would be reversed. But regardless, since the first three items directly affect flight safety, they are more critical than the last two.

Adequate crosswind handling not only designs the control surfaces as will be shown in the next chapter, but also directly affects flight safety near the ground—the most critical flight regime. In similar fashion, rate of climb has a direct bearing upon the level of flight safety offered by any airplane. This is evidenced by the ability or inability to climb out of short fields or over hills bordering an airport or lake; ceiling capability in mountainous or icing conditions; missed approach climb response; and hot day operation—to mention a few conditions. Criteria 1 and 2 obviously have a major effect upon safety.

Criteria 1 and 3 combine to provide an easily handled airplane, which also directly relates to safe operation. Responsive controls and inherent stability about all three axes are essential for IFR work. Fortunately, meeting criterion 3 also results in an airplane that is pleasant to fly in stable air and one that does not become a demon in turbulent conditions.

Useful load is always marginal, so the degree of inadequacy is really the determining factor for criterion 4. However, it would help if aircraft advertised to carry four adults could carry four 170-lb people plus 100 lb of baggage in addition to full fuel, oil, primary blind flight instruments, and basic IFR radio equipment. This would require a total useful load of about 1130 lb. As of this writing, no production 150-hp four-place airplane can carry that weight. The Cessna 172 Skyhawk and the Piper PA28-151 Warrior come closest to the target, and, not surprisingly, enjoy the largest production volume of the fixed gear, 150-hp class. However, at gross weight on a hot day their climb performance leaves something to be desired. The 180-hp versions of each materially improve this important requirement while also slightly increasing useful load and cruise—and considerably increasing cost.

On the surface, high-speed cruise performance for private aircraft seems really critical only if frequent trips are to be made between distant locations—say over 300 statute mi apart. After all, the ability to fly at 100 mph in a straight line usually more than doubles automobile cruising speed, since major highways seldom connect desired destinations by direct routes. Of course, winds aloft do affect ground speed and always seem to be blowing from the desired heading. In view of this, the higher the *economical* cruising speed the better. Because no one enjoys listening to a 2500-rpm power setting for hours on end, the ability to cruise small aircraft at 130 mph or faster at 2300 rpm or less can be most beneficial from both the sound level and fuel economy standpoints. So there is much merit to having a relatively high cruising speed at fairly low power. A brief ground speed example illustrates this quite clearly: With a rather frequently experienced 30-mph headwind condition, a 110-mph cruising speed delivers only 80-mph ground speed; however, a 150-mph cruise will result in a 120-mph ground track—a 50 percent increase in distance covered for a 36 percent increase in cruising speed. So design cleanliness is important even for small private aircraft, and particularly so when fuel prices seem to increase faster than performance.

Now how does the private aircraft designer achieve all these goals? Some are in direct opposition, for example, high cruising speed vs. low approach and landing speeds. High cruise usually means low wing area as well as plenty of power, while low stalling speed requires much wing area and/or low gross weight. But with high power we have an expensive engine and a heavier one, adding to cost, weight, and landing speed.

If the airframe is too heavy, the useful load will be too low to be practical; if too much wing area is sacrificed to obtain high cruising speed, both short-field capability and climb will be compromised; if too much power is used to obtain climb and cruise for a given gross weight, the useful load will go down as fast as the cost goes up; if too much wing area is added to obtain short-field capability, the cruising speed will decrease while the weight and cost increase.

In view of the above it is not surprising that designers study similar successful aircraft before starting upon any new project. Compromise is obviously required, with trade-off decisions determining the level of acceptance for the resulting airplane. In order to define and describe aircraft design procedures, much of this book will be concerned with just how such decisions are reached along with the effect of various options upon aircraft performance and handling characteristics.

✳   ✳   ✳

These introductory chapters have provided background information relating to all general aviation aircraft. We shall now consider the practical application of design requirements for *private aircraft* beginning with particular emphasis upon criteria 1 through 5 of this chapter, which will be presented in that order, with two chapters devoted to criterion 3. To be more precise, we shall be reviewing the effect of various design criteria upon the handling and performance characteristics required by small aircraft. We shall also consider why these aircraft must look as they do to satisfy FAA regulations—and, more important, to be accepted by private pilots. For it is they who determine whether or not a new airplane will be successful.

In final analysis, it is the customer who really designs the airplane.

REFERENCE   4.1   FAA FAR Part 1: Definitions and Abbreviations, p. 1.

5 CROSSWIND HANDLING

DESIGN REQUIREMENTS

Crosswind conditions usually affect the beginning and end of every flight. Although the landing phase is generally more critical, things can get quite active during takeoff if the rudder and ailerons are not effective at low speed. This situation can be bad on land and may be dangerous on water, so responsive controls are essential for satisfactory crosswind operation.

When near the runway surface, it is most reassuring to have excess aileron control available to handle crosswind gusts and changes in wind direction. At low speeds, ample aileron power during flare can correct momentary trim changes in roll and yaw, while marginal or insufficient roll control may result in premature ground contact plus accompanying bounce, ground loop, or wing-tip damage.

If the ailerons are too small in chord or span, when deflected downward they cannot generate enough increase in life coefficient (through change in wing-section camber) to provide the lift required to raise a low wing. Conversely, when deflected upward small ailerons cannot reduce lift to the degree necessary to rapidly lower a high wing. Since the ailerons are interconnected and act together to provide lateral control, it is obvious why weak ailerons cause a marginal rate of roll and could make an airplane impossible to handle in gusty crosswind conditions.

To be effective in changing wing camber, an aileron should have a chord aft of the hinge line of 15 to 25 percent of the wing chord, and a total area (both ailerons) of about 7 to 10 percent of the total wing area as shown by Fig. 5-1a. Of course, too much area will make the ailerons feel "heavy" to operate. In addition, the outer end of the aileron should be at or just inboard of the wing tip. This location places the center of aileron area and lift as far from the airplane centerline as possible to provide maximum rolling moment about the longitudinal axis. As a note of interest, increasing wing area by adding span outboard of the ailerons effectively buries them, resulting in an undesirable reduction in roll rate and control response.

To further increase roll capability, the wing panel may be tapered from root

FIGURE 5-1
(a) Aileron area and power.
(b) Dihedral angle.

(1) Aircraft	(2) Wing area S_W sq ft	(3) Aileron area $S_{A_{total}}$ sq ft	(4) $\dfrac{S_{A_{total}}}{S_W}$	(5) Airplane to center of aileron area, ft	(6) Aileron power (3) × (5)
SIAI Marchetti SF 260 (260 hp)	108.5	8.50	.078	11	93.5
Czechoslovakia Zlin Z-526L (200 hp)	166.3	14.8	.089	13.2	195.4
Wassmer Super 4/21 (235 hp)	172	13.35	.078	13	173.5
Jurca Sirocco MJ-5H2 (160 hp)	108	10.0	.093	9	90.0
Cessna Cardinal (180 hp)	174	18.9	.108	12.9	243.8
Piper PA28-151 Warrior (150 hp)	170	16.1	.095	12.4	199.6
Piper PA28-200R (200 hp)	170	12.4	.073	11.5	142.8
Robin HR 100 (180 hp)	176.5	13.9	.079	13	180.7
Thurston Teal (150 hp)	157	12.8	.081	13.3	170.2
Sequoia S285 (285 hp)	130	13.8	.106	11.9	164.2
			.088	Average	165.4

(a)

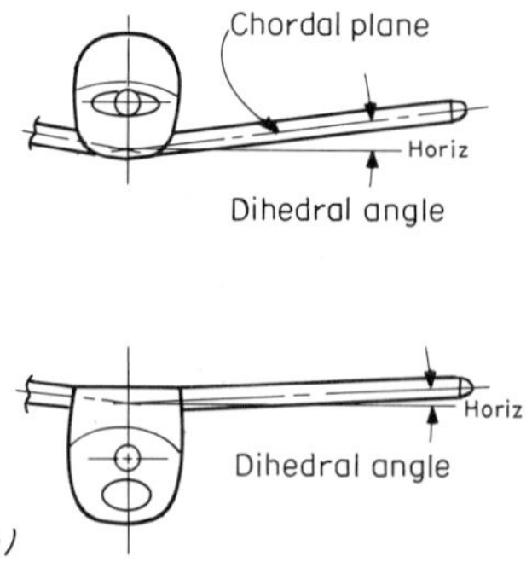

Chordal plane

Horiz

Dihedral angle

Horiz

Dihedral angle

(b)

to tip or possibly only in the outer portion of the span ahead of the ailerons. This reduction in outboard area compared with a rectangular wing decreases wing panel damping action and so makes the ailerons more powerful for a given aileron area and spanwise location. By way of illustration, we all know that a small paddle may be moved through the air more easily than a large one.

To improve approach and landing control, adverse yaw must be kept to a minimum. As you will recall, adverse yaw is the tendency of the airplane to turn to the right when left aileron is applied, and to the left when using right aileron. Adverse yaw is caused by the increased camber drag of the down aileron and reduced camber drag of the up aileron. It is usually most pronounced and trying at relatively low airspeeds occurring at corrective maneuvers during the landing flare. Adverse yaw is particularly magnified at lower speeds because fin and rudder effectiveness decrease rapidly with airspeed; in a falling-dominoes-type reaction, reduced fin effectiveness causes a loss of directional stability and, therefore, the ability to hold a desired heading by correcting for adverse yaw.

Since decreasing control response is obviously undesirable at this critical point in the landing procedure, control effectiveness must be improved to the extent possible through modified aileron and vertical tail design. The Frise aileron and associated variations of this type have a section forward of the hinge line which projects down into the airstream below the wing when the aileron is raised. This action has the aerodynamic effect of dragging back the wing panel having an up aileron, thereby turning the airplane into the direction of bank called for by aileron deflection. Most aircraft have adaptations of this type of aileron design (Ref. 5.1), as may be noted during your next preflight inspection.

No discussion of roll control would be complete without noting the effect of *dihedral*—a full subject in itself. Dihedral is the spanwise angle the wing chordal plane makes with the horizontal, and is usually greater for low-wing aircraft than for high-wing designs as shown by Fig. 5-1*b*. A larger dihedral angle is necessary for low-wing aircraft to bring the center of lift closer to or above the center of gravity. This location increases *static stability* about all three axes by having the lift force act above the airplane weight centered at the center of gravity. This effect is clearly shown in Fig. 5-2 as well as by Figs. 2-4 and 3-4. The center of gravity is below the lift center of high-wing aircraft, providing a form of built-in pendulum stability which assists in righting the airplane in roll and pitch. Such corrective action is a measure of static stability, which may be considered as the ability of an airplane ro return to trim attitude when displaced from level flight. The faster and more complete the return, the more stable the airplane; usually, the further the center of gravity is located below the center of lift the more stable an airplane will be in cruising flight. This is a major and distinct advantage of high-wing aircraft.

While dihedral is related to both the lateral and directional stability characteristics of the airplane, too much dihedral will make precise control of the degree of roll most difficult in gusty conditions. This is so basically because the

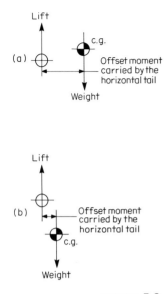

FIGURE 5-2
The effect of lift and weight forces upon static stability. (*a*) Low wing position. A relatively unstable condition; the corrective action of the horizontal tail prevents the lift and weight forces from rotating over 90 degrees to become vertically oriented with the center of gravity directly below the center of lift. (*b*) High wing position. A relatively stable relationship requiring very little horizontal tail load to balance the few degrees of angular rotation that would align the lift and weight forces. This force relationship provides pendulum stability as explained in the text.

down (windward) wing has more projected area and provides more lift about the airplane centerline than does the raised (downwind) wing. As a result, more aileron power must be available to keep the windward wing down on any design having too much dihedral. For crosswind control, moderate dihedral is required for stability, with somewhat more aileron power necessary for low-wing designs.

To satisfy these requirements we find our most responsive crosswind aircraft fitted with generous, but not overly heavy feeling Frise-type ailerons positioned as far out on each tapered wing panel as practical. And not surprisingly, this configuration is found on all of today's most popular private aircraft.

As previously noted, the level of directional stability provided by the fin and the amount of rudder control available will both decrease with airspeed. In fact, if approaches are made with plenty of flap down and decreasing speed, sometime during the flare rudder control will usually be found in short supply—and probably before the airplane runs out of roll control. This may be due to partial blanketing of the vertical tail by the wing and flaps, or possibly from turbulent flow off the cabin structure. Fortunately, the slipstream effect from application of power usually restores directional control quite rapidly.

The use of a T-tail configuration not only makes the fin and rudder more effective due to the end plate action of the stabilizer location, but also positions the horizontal tail above wing turbulence. The addition of a dorsal fin will tend to straighten turbulent flow over the fin and rudder, increasing the effectiveness of both while providing additional fixed fin area to improve directional stability at low speed. The new Beechcraft model 76 twin of Fig. 5-3 incorporates both of these features.

The so-called uncontrollable deep stall experienced by heavy transport and military aircraft has been frequently blamed upon the T-tail configuration; actually high-wing loadings and powerful flaps are the real cause due to the resulting strong downwash flow off the wing and over the high stabilizer. Deep stall should not be experienced by light aircraft having a T-tail design since relatively low-wing loadings cannot establish a downwash flow of sufficient energy to overcome pilot effort and thereby lock the elevator in the up position. A few designs are now flying with a dorsal fin and T-tail combination, and we will see this arrangement become more standard in the future.

Readers interested in tail surface design will find acceptable criteria and related discussion presented in Chap. 7.

The crosswind landing gear, further described in Chap. 14, was an interesting post-World War II effort to solve some of the more serious crosswind problems that developed as grass airfields were replaced by asphalt runways. A number of different systems were patented, varying principally in caster design. This improvement was intended for conventional gear aircraft having the main wheels forward plus a tail wheel. The tail wheel was steered by the rudder and was usually designed to swivel free after turning 30° to the right or left.

Crosswind landing gear, also known as "self-castering" gear, was thoroughly demonstrated to the industry during the mid- and late-forties and flown by many pilots at all levels of proficiency. This unique main gear design permitted aircraft to land with a crabbed heading and then run out in a skewed fashion with the airplane pointed one way and the wheels another, as shown in Fig. 14-3. This was accomplished automatically at ground contact by having the main wheel axles "break free" under a side loading of predetermined value.

Unfortunately, the gear produced some odd effects such as slewing off the runway if a turn were started at a speed high enough to permit castering. A few of these negative handling characteristics plus rapid acceptance of tricycle landing gear terminated further development of crosswind gear during the late 1940s.

At least one very promising approach to solving the flight control coordination problem was found deficient and dropped primarily because of crosswind operational requirements. Shortly after World War II the lightplane industry was vigorously pursuing the so-called two-control airplane. As you will recall, a two-control airplane has interconnected ailerons and rudder plus normal elevator control. This type is best remembered for the still active Ercoupe designed by Fred Weick just before the war, and produced in great numbers after 1946. However, as time passed the crosswind shortcomings of two-control aircraft in the hands of the average pilot became all too obvious. It was not possible to bank into a stiff crosswind to check drift without also flying off the runway heading—and any crab angle could not be straightened at touchdown in an airplane without independent rudder control. Fred Weick says his solution was

FIGURE 5-3
Beechcraft model 76 light twin
with T tail
Beech Aircraft Photo

to bank into the desired heading, level the wings, and land at the flight crab angle. He never had a landing gear failure in his Ercoupe; so this may be an acceptable crosswind technique for all two-control aircraft, provided the landing struts can take the side loading.

Although some two-control aircraft were a delight to fly, producing ball-centered turns with an automotive-type steering wheel (so anyone who could drive could also expect to fly with minimum practice—or so the ads declared), two-control aircraft became a real handful for many during approaches in a moderate crosswind. As a result, serious attention returned to the three-control system of ailerons, elevator, and rudder which is now universal for private aircraft.

Having reviewed the effect of control surfaces upon handling response and flight characteristics, let us now run through a set of crosswind operations to see just what happens when we apply these controls in flight.

WHEN THE CROSSWINDS BLOW

Many of us were first attracted to flying when exposed to the challenge of mastering flight, possibly as a passenger in a friend's airplane or during the thrill of an introductory ride. And there is no better way to keep this mastery current than by practicing crosswind takeoffs, approaches, and landings.

All too many pilots, particularly those who only fly a few hours each month, retreat to "hangar flying" as soon as the surface wind gets much over 15 mph or approaches an angle of 40° to the runway. Flying is not a particularly difficult art, but it does require practice and acquired skill—especially in adverse or less than ideal conditions.

We tend to shy away from crosswind flying for many reasons. If the winds are high, the air is likely to be gusty and turbulent due to frontal passage, and the wind angle will vary all over the field along with the gust velocity. Certainly not a relaxing situation and one easy to sit out, particularly when the tower reports winds leading into the danger zone of Fig. 5-4. We all know that a 30-mph wind at 60° to our runway requires correction on downwind and base, but we are quite far from the ground then and feel relatively safe. The conditions seem to magnify as the runway approaches (or drifts to one side) during final. But the time to have become alert was during the early phase of the downwind leg when correction should have been applied to track (fly over the ground) parallel to the runway, and so get tuned up for the base and final legs.

Some pilots have been advised not to use flaps during fairly high or gusty wind conditions. Flaps can and should be used on all approaches since they act to stabilize most aircraft in both pitch and roll, as well as reduce ground contact speed. However, the degree of flap application will vary with the airplane type and with surface conditions and so presents another decision problem for many pilots.

Ground handling during taxi and takeoff also becomes an increasingly demanding problem in high winds, and in fact may require some controlled rate

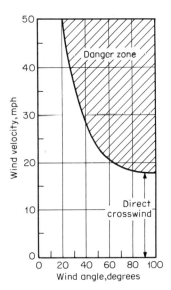

FIGURE 5-4
Maximum safe crosswind velocities

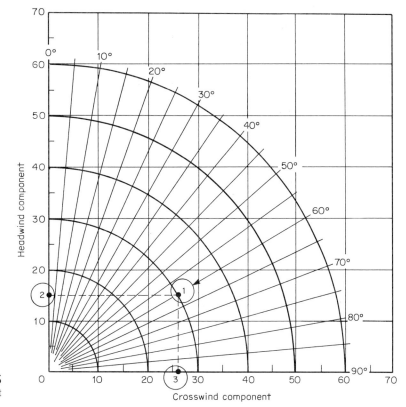

FIGURE 5-5
Wind component chart

CROSSWIND OPERATION

of reduction of aileron displacement during takeoff in heavy crosswind conditions; no test of skill, but frequent practice would help. Since most pilots are concerned about their ability to handle adequately that airplane they must sign for, let us take a further and closer look at the much magnified and frequently maligned crosswind operational requirements.

As shown by Fig. 5-5, any crosswind may be divided into two components that act upon the airplane; one counter to the direction of flight, acting at a lower velocity than the surface wind, and a second component at 90° to the direction of airplane motion. The first component reduces ground contact speed similar to any headwind, while the right angle component tries to make the airplane drift, weathercock, or flip over—either singularly or all at once.

For example, with a surface wind velocity of 30 mph blowing at a 60° angle to the runway heading (Point 1 of Fig. 5-5), the effective headwind reducing ground speed will be 15 mph (Point 2), but the 90° crosswind component trying our skill will be 26 mph (Point 3). Small wonder these conditions lie within the danger (alert) zone of Fig. 5-4.

We have a few tools to combat such conditions, and with practice we can operate in fairly severe weather as though on a normal flight, provided heavy crosswinds are never treated as a routine matter. In fact, it is strongly recommended that any initial practice in severe crosswind conditions be under the guidance of a competent instructor—preferably sitting in the right seat.

Although all controls participate fully in crosswind work, *rudder* plus *aileron* coordination really come into play first when setting up the crab angle to hold course on downwind, base, and final. There is no need to crank in upwind aileron 800 ft above the airport on final; why develop a shaking set of calf muscles from holding opposite rudder to keep centered on the runway for 2 min or more? Further, during all this time the airplane will have crossed controls so any sudden drop in gust velocity or a sharp upgust, reduction in approach speed due to inattention, or turbulence-induced trim change will verge upon spin conditions. Certainly a spectacularly unpleasant way to end any flight.

A much better practice is to keep the runway centered in your sights by maintaining a comfortable crab angled approach, power off or low, coupled with a steady descent speed and rate controlled by the *elevator*. About 50 ft above the runway *aileron* is applied to drop the upwind wing to stop drift while at the same time *opposite rudder* is used to keep the nose straight and the runway centerline boresighted. Apply more or less rudder and aileron as necessary to check immediately any runway centerline drift downwind or upwind as gust conditions vary.

Commence to flare and follow through to land on the upwind main wheel, then the downwind main wheel, and finally the nosewheel or tail wheel. Before contacting with an interconnected, steerable nosewheel it is advisable to get the rudder centered or you may go shooting off the runway downwind (due to holding opposite rudder which casters the nosewheel downwind); however, upwind aileron should be held during the runout.

An alternative crosswind approach procedure holds the upwind crab angle during flare, with a final downwind rudder kick used to straighten everything out just before touchdown. I was taught to fly my PT-19 this way, and it is great sport and truly a test of skill to execute this correction maneuver properly and at the precise moment required. But only if the wind is light and landings are made upon a nice grassy surface. When operating in a high crosswind and landing onto an asphalt or concrete runway, watch out!

If the crab is not fully corrected, the combination of upwind facing nosewheel plus ground speed could ground-loop a tricycle gear airplane as surely as a tail dragger. And if the crab correction is kicked out too high above the runway, downwind drift will thoroughly test the side load capability of the main gear struts. As a final insult, too much downwind rudder correction at touchdown can result in running away downwind. These maneuvers are described in more detail in Chap. 14, Landing Gear Design.

For the average pilot, crosswind conditions are more easily handled with tricycle landing gear, preferably with the spanwise distance between wheels, known as the tread, approaching one quarter the wingspan. The wider the tread, the greater the angle of bank possible to handle crosswinds before dragging a wing tip. While a high-wing design might appear capable of steeper banks and so be able to operate in more severe crosswind velocities than would be possible with a low-wing type, this does not seem to be the case nor a critical consideration since low-wing designs normally have the main gear located well out on the wing. Either configuration is equally effective in crosswinds provided sufficient approach and flare speed is maintained to assure aerodynamic control.

When flying speed is too low for the crosswind conditions, the airplane will soon let you know by running out of rudder and aileron control. When this happens, either increase speed to regain control or apply power and go around again; running out of rudder or aileron can be an eerie and long remembered experience.

Recommended approach and primary flight control procedures have purposely been discussed separately from flap operation. Flaps should be used to some degree during final approach and possibly when on base leg, depending upon height and speed in the landing pattern, wind strength, and runway length available.

Just how much flap to apply is a matter of judgment influenced by airplane handling characteristics as well as by surface wind velocity and direction. The use of 10 to 15° of flap to improve stability during the early portion of the final leg, and increasing to 20 to 30° during the final half would seem a good compromise in crosswinds blowing up to 20 mph and angled up to 60° with the runway. Possibly more flap could be used for shorter runways—a few practice landings should provide the answer for you. But remember to keep flying speed higher than usual with flap operation in order to maintain an adequate margin of control. Since the wind is blowing harder than normal, the headwind component will reduce ground contact speed. Also, the extra flying speed is most helpful when a gust drops 10 to 15 mph or changes heading 30°. At any rate, full flap deflection would seem excessive under extreme crosswind conditions. Also, *be sure to dump the flaps once finally on the ground;* the low contact speed obtained by using flaps assures that any normal gust cannot lift a wing once the flaps have been retracted.

Floatplanes, flying boats, and amphibians have somewhat special problems during high wind conditions. When landing on the average waterway, any seaplane can usually select an approach path heading directly into the wind and execute a full stall landing at relatively low surface contact speed. However, when water approaches must be made in crosswind conditions due to a narrow canal or creek, the amount of crosswind that can be handled is somewhat limited by wing floats in the case of flying boats and amphibians, and equally so for float seaplanes by the prospect of tripping the upwind float on wave tops.

Hull-type amphibians are also limited by possible wing float contact with the runway surface when operating in crosswinds on land, but this is not particularly serious since the landing gear is (hopefully) extended and provides additional ground clearance.

As a final note, all crosswind approaches and landings are frequently made increasingly difficult by the presence of upwind trees, buildings, embankments, snowbanks, and other obstructions which create both additional turbulence and rapidly variable wind headings. This same effect can be noted to a lesser degree during final approach when course correction may be required to accommodate variations in wind direction and velocity with altitude—as any IFR pilot knows only too well from trying to keep the localizer needle centered during the final moments of a long approach.

$$* \qquad * \qquad *$$

Acceptable crosswind handling characteristics can be obtained through various design features, practically all of which are incorporated in current production aircraft as listed below. In summary, crosswind takeoffs and landings can be readily performed by the average private pilot if the airplane design has:

1. Tricycle landing gear

2. Aileron, rudder, and elevator systems that are responsive at low speed

3. A combination of wing area and flaps adequate to provide low landing and takeoff speeds (this is particularly important for short-field and seaplane operation)

4. A responsive engine and throttle system for control in gusty air or for climb-out when a landing must be aborted

5. Basically stable flight characteristics

This chapter has shown why these features are important and what design parameters are required to make them effective. In order to satisfy these criteria, the most popular aircraft quite understandably look somewhat alike. For the near future, the modern private airplane will have a tapered wing, T tail, dorsal fin, effective ailerons, and tricycle landing gear with a relatively free-to-caster nosewheel. Interestingly enough, the design of all these components will have been influenced by crosswind operation requirements and related FAA flight demonstration criteria.

REFERENCE   5.1   Fred E. Weick and Richard W. Noyes, *Wind Tunnel Research Comparing Lateral Control Devices, Particularly at High Angles of Attack. II—Slotted Ailerons and Frise Ailerons.* NACA Report No. 422, 1932.

6 TAKEOFF AND CLIMB

In the preceding chapter we have seen how the primary control surfaces interact to provide acceptable crosswind handling characteristics. In similar but more complex fashion the powerplant, wing area, control surfaces, fuselage, landing gear, and design weight must be optimized and combined to provide the best possible performance during all phases of flight. As we now get into the fundamentals of airplane design, let us first review the ground handling, takeoff, and climb characteristics necessary for pilot and FAA acceptance of current production aircraft.

GETTING UNDERWAY
While not initially part of the takeoff sequence, ground handling is of prime importance for aircraft acceptance. If an airplane seems difficult to taxi away from the ramp or tie-down area, any pilot will naturally feel a bit apprehensive and annoyed before the flight begins. This is no way to enjoy flying—or to sell airplanes either, for that matter.

The type of landing gear is most critical here, and there is no doubt that a wide tread, tricycle configuration is much easier to handle than the so-called conventional arrangement of main wheels forward plus a tail wheel. The addition of a steerable nosewheel and differential braking complete the features necessary to make tricycle landing gear the easiest to control under all ground handling conditions, including takeoff and landing runs.

In addition to these control benefits, tricycle gear positions the airplane in a fairly level attitude which improves forward visibility, further adding to pilot and passenger safety as well as improved ground handling. In view of all this, no wonder the most popular late-model aircraft are equipped with tricycle gear, a steerable nosewheel, and independent main wheel brakes.

TAKING OFF
As the takeoff run gets underway there is nothing so satisfying as pushing the throttle home and feeling the seat back give your spine a good shove. This is

acceleration at work and the amount available can be very critical for operation out of short fields or ponds. Although we seldom experience takeoff acceleration exceeding 1/3 g with any of our small aircraft (for example, 800 lb of thrust at 2400 lb takeoff gross weight = 800/2400 = 0.33 g), even this value will be difficult to obtain unless the airplane has:

1. A constant speed propeller or a fixed pitch type designed for maximum climb thrust
2. A weight to power ratio of 12 or less ($W/hp < 12$), which means
3. Relatively low gross weight for the power available

In analyzing acceleration, the initial takeoff parameter, we find that propeller design, powerplant size, airframe weight, and useful load are all critical items. A variation in any one of these will affect takeoff distance favorably or unfavorably.

Since thrust is a function of blade design and pitch angle, engine power, ambient temperature, and altitude, our source of accelerating force is obviously sensitive to many things. Reduced thrust delivered by a fixed pitch propeller in comparison with a constant speed type means lower acceleration and an increase in time and distance to reach takeoff speed. This occurs because at takeoff a constant speed propeller will have a lower blade angle than a fixed pitch prop, and so deliver greater thrust for increased acceleration. The fixed pitch propeller blade angle is normally set for cruise flight, which means the blade pitch would be greater than optimum for takeoff. This is necessary since a fixed pitch propeller designed for maximum thrust at takeoff would overrev at cruising speed due to blade unloading, as explained at greater length in Chap. 11.

Axiomatically, takeoff time and distance will be decreased by replacing a fixed pitch cruise propeller with a constant speed one of correct design.

Increased thrust and improved takeoff acceleration will be realized from added horsepower, provided the propeller is changed or modified to accept increased engine torque. Again we have a trade-off, however, since more powerful engines usually are heavier engines and either the gross weight will increase—which tends to lower acceleration—or useful load must be reduced to maintain the original gross weight. In addition, airplane cost and fuel consumption will both increase.

Conversely, more weight at constant power reduces acceleration as shown by our simple formula:

$$\text{acceleration} = \frac{\text{thrust}}{\text{gross weight}}$$

This means airframe weight and useful load have a direct bearing upon takeoff time and distance, and explains why you must always be aware of takeoff weight vs. airplane design gross weight. Overloading may mean flying into the trees at the end of the runway instead of over them.

As we have been taught, temperature and altitude conditions may also result in a tree-studded takeoff. In fact, temperature and altitude combine to form *density altitude* which has direct effect upon engine power, takeoff thrust, and takeoff speed over the ground. An airplane will takeoff (or stall) at the same indicated airspeed regardless of air density, since lift includes the dynamic pressure relationship

$$q = \rho \, \frac{V^2}{2}$$

where q = dynamic or impact pressure as read by the Pitot tube
ρ = air density (affected by altitude and temperature)
V = airplane velocity

By way of explanation, for a given wing area, airfoil section, and angle of attack, the lift developed depends upon the mass flow of air passing over the wings. Since this mass flow must always have the same total value regardless of altitude or temperature when the same load is to be lifted by the same airplane, it follows from the above formula that as air density decreases, airplane velocity must increase to keep q constant. Simply interpreted, this means increased takeoff ground speed and distance as air density decreases. Of course, engine and propeller output also decrease with reduced air density. The additive combination of lower power (which means lower acceleration) and higher ground speed necessary to develop the lift required can easily exceed the takeoff distance available, again ending the takeoff in the trees—but this time due to density altitude rather than excess weight.

For pilot convenience the effects of temperature and altitude upon takeoff and climb have been put into a simple circular calculator known as the Denalt performance computer. It may sound as though this aid was developed by a French pilot or engineer, but the name is actually a combination of "density altitude" contracted into the word *Denalt.* Separate Denalts are available for fixed pitch and for variable pitch propellers, and both may be obtained from the Superintendent of Documents, U.S. Government Printing Office, Washington, D.C. 20402.

The Denalt performance computer includes takeoff distance variations with density altitude as well as climb rate, and should be part of your flight kit. Since takeoff distance and climb rate are important safety considerations, note that on a 60°F day an airplane equipped with a fixed pitch propeller will require 1.9 times the takeoff run at 4000-ft pressure altitude that it does at sea level. And if the climb rate was 700 ft/min at sea level it will decrease to 460 ft/min at 4000 ft and 60°F. At 90°F the 4000-ft-altitude takeoff becomes 2.1 times the length at sea level, and the climb further reduces to 365 ft/min. So beware of high temperatures and altitudes at takeoff; not only is the ground run increased, but climb-out over local obstacles is severely decreased as well.

Terrain may affect takeoff as much as density altitude. If we use a smooth asphalt or concrete runway as a standard of comparison, anything else is downhill. For example, a dry, grassy runway will increase takeoff and 50-ft obstacle clearance distances by 7 to 10 percent, while long grass nearly as tall as hay may restrict acceleration to such a degree that takeoff is impossible.

Snow and slush are also culprits. In addition to making runways slippery and dangerous, these conditions can freeze wheels and brakes and cause jet flameouts from slush ingested during takeoff. Airline flight standards presently prohibit taking off from runways covered with 2 in. or more of slush. And just consider the size of their wheels and thrust-to-weight ratios compared with our small private aircraft; if they have trouble, light aircraft will fare even worse. Blown snowdrifts and hidden ice ridges can be as dangerous to a landplane on takeoff as a submerged sandbar is to a seaplane. There is no design solution for these conditions, nor is there useful advice for that matter, other than recommending the installation of snow skis in place of wheels.

By this time it must be quite apparent that density altitude and terrain effects can combine into a potentially dangerous set of takeoff conditions. This is particularly true for operation from grassy or snow-covered upland meadows, and partially explains why this type of flying requires experience. The other part of the explanation may be found in local wind conditions due to mountain ridges and slopes, uphill takeoff patterns necessary to clear terrain, and local obstacles that create turbulent flow.

Aircraft with conventional landing gear are best suited for these conditions because the tail wheel may be raised early in the takeoff run and the airplane then held at a trim angle producing minimum drag during takeoff. The Piper SuperCub is a classic example of such design and short-field capability.

Unfortunately, the relatively short wheelbase of tricycle gear aircraft will cause them to hobbyhorse over uneven ground as the nosewheel surveys every contour. This results in a constantly changing wing angle of attack, preventing the buildup of steadily increasing wing lift. Because of the location of the main wheels on tricycle designs, considerable airspeed is required before the horizontal tail can develop enough down load to rotate and lift the nosewheel clear of the ground as shown in Fig. 6-1a. Further, this down load is added to the wing load per the discussion of Chap. 2 and Fig. 2-4. Of course this increases takeoff speed since more wing lift must be developed. The speed required to lift a tail wheel is usually lower than nosewheel lift-off speed because a tail wheel is normally more lightly loaded than a nosewheel. In addition, the distance between the main gear and tail is greater, requiring less tail load to rotate a given gross weight as indicated by Fig. 6-1b.

Also note that this lift will be directed *upward*, reducing, rather than adding to, wing lift required for takeoff. And this is the basic explanation of why tail-wheel aircraft are preferred for short-field and bush-country operation; due to their landing gear arrangement, tail-wheel aircraft are able to take off at lower speed and in less distance than tricycle gear aircraft. Once flying speed is

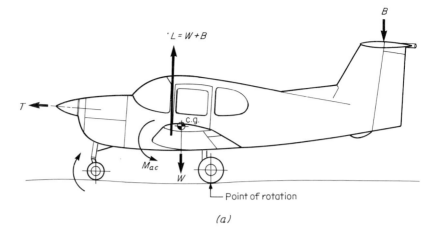

$$L = W + B$$

(a)

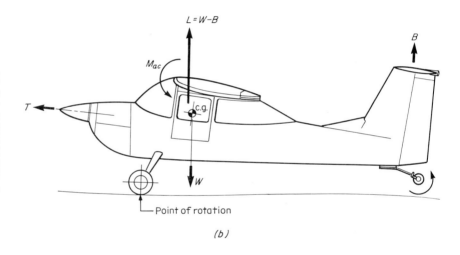

$$L = W - B$$

(b)

FIGURE 6-1
Tail loadings required to raise
auxiliary wheels clear of terrain. (a) Horizontal tail action
required to raise a tricycle
gear nosewheel adds to total
wing loading. (b) Horizontal
tail action required to raise a
conventional gear tailwheel
reduces total wing loading.

reached it will be necessary to rotate the airplane about the main wheels to increase angle of attack and so develop the lift required to unstick and climb out
of ground effect. This trim change results in a tail down load for tricycle gear
aircraft and will require lowering the aft fuselage of tail-wheel types. Both actions tend to increase the wing loading and to some extent prolong the takeoff
run.

In an effort to minimize this effect for tricycle gear aircraft, some designers
set the wing at such an angle to the ground line that the airplane can effectively
fly off in a three-point attitude. This goal can only be realized on a smooth
runway as previously noted. In addition, the three-point attitude will generally
develop more drag due to increased projected frontal area, so in reality the
takeoff run may be increased. The Cherokee series approaches this arrangement which has the further objection of causing nosewheel contact damage

when the landing speed is too high. The accompanying nose-low flight attitude (frequently referred to as *wheelbarrowing*) can subject the nosewheel to overloading and structural failure.

Of course, tail-wheel aircraft are designed to stall in the three-point attitude and may be taken off in that fashion, although the added frontal area drag should again be expected to increase the run beyond that of an early "tail up" takeoff technique.

Due to hydrodynamic suction effects, breakout speed is usually well above stall speed for hull seaplanes and to a lesser degree for floatplanes. This explains why a seaplane seems to pop up out of the water once it is ready to take off; the surplus speed is converted into a short burst of excess lift at the takeoff trim attitude.

With landplane, seaplane, tricycle, or conventional (tail-wheel) gear, takeoff speed should be at or near 1.3 to 1.4 times the stalling speed, with the higher value preferred for rough air conditions. As you may know, current FAA design specifications prohibit a gross weight stalling speed in excess of 61 knots (70 mph) for private single-engine aircraft (Ref. 6.1). Fortunately, most small airplanes stall around 60 mph (52 knots), so our takeoff speed should never have to exceed 85 mph (73 knots), even in turbulent conditions.

As a further consideration, FAR 23 requires sufficient elevator power to lift a tail wheel at or before $0.80\ V_{s_1}$ or a nosewheel at or before reaching $0.85\ V_{s_1}$ where V_{s_1} is the stalling speed in the takeoff configuration (Ref. 6.2). As a practical matter, in order to properly handle rough terrain it should be possible to lift either auxiliary wheel at a considerably lower airspeed; most aircraft can do this quite handily.

IMPROVING TAKEOFF   So much for the fundamentals of takeoff, which are basically determined by the simplest laws of physics. Fortunately, we have developed ways to improve takeoff performance beyond these limitations by decreasing the takeoff distance while also increasing the rate of climb. Increased horsepower will do this, but at the price of useful load, greater cost, and reduced economy as previously noted. A much more effective way is the sound engineering approach of solving a problem at its source, in this case, by lowering the takeoff speed.

The lower the end speed necessary to rotate and climb out, the shorter the takeoff run for a given power and weight condition.

This is so because distance covered over the ground is a function of velocity and time; if takeoff velocity is reduced but acceleration remains unchanged, the distance required to reach this lower speed will also be reduced. In addition, lower takeoff speed means the available acceleration will be applied for a shorter time; and so we have a double gain in that both time and distance are reduced proportionately with takeoff speed. And I'm sure you will be pleased to know this now completes our takeoff physics lessons.

Of course, lower takeoff speed permits operation from shorter runways and smaller ponds—or safer operation from larger ones, and with a steeper climb angle as well. Reduction of takeoff (and landing) speed is particularly important for seaplane design since hull strength requirements are based upon a function of stalling speed squared. As an illustration, reducing the stalling speed 8 percent from 65 knots to 60 knots lowers the hull bottom loading by 17 percent.

There are basically three ways to reduce takeoff speed for a given gross weight:

1. Increased wing area

2. A more advanced airfoil section having higher lift coefficient values

3. Wing flaps

None of these represents an optimum solution to the problem in itself. As we review each method the necessary trade-offs provide an excellent lesson in basic engineering, for practically all phases of aircraft design require compromises favoring operational requirements. In this case, successful compromise may mean the use of all three methods to improve takeoff and landing performance.

The use of increased wing area reduces landing speed in inverse ratio to the square root of the wing area change, as shown by the formula

$$V_{s_2} = V_{s_1} \sqrt{\frac{S_{w_1}}{S_{w_1}}}$$

This means that, for a given gross weight and airfoil section, doubling wing area would lower an 80-mph stalling speed only down to 56.6 mph—not 40 mph. As might be expected, any such wing area increase would change more than just the wing weight since more tail area would be necessary to satisfy stability requirements, the fuselage weight would go up to carry the greater loads, etc. In fact, the remaining useful load would soon drop below an acceptable value.

For practical design an increase in wing area can be used to provide some reduction in stall and landing speeds, but if considerable change is required the designer must seek other solutions—one of which is an increase in lift coefficient.

If the calculated stalling speed is quite high due to selection of a low lift airfoil, there are many other wing sections available that can deliver increased lift without any adverse side effects. While the tendency is then to select a very high lift airfoil, remember that any accompanying high section moment coefficient means an increase in tail loading and structural weight (as shown in Chap. 2, Level Flight Loads); this should be avoided.

For current production private aircraft, a basic lift coefficient value approaching 1.3 to 1.4 would not be unusual as shown by Fig. 6-2. While the newer GA(W)-1 and GA(W)-2 airfoils developed by NASA may exceed these

Airfoil section	C_{Lmax} @ RN = 3×10^6 *	α Wing section angle of attack @ C_{Lmax}†	C_{Mac} Wing section moment coefficient	C_{domin} Wing section drag coefficient @ Std. roughness	Cruise C_L range @ C_{domin}	$\left\{\dfrac{Percent}{thickness} = \dfrac{thickness}{chord\ length} \times 100\right\}$
4415	1.42	12°	−.092	.0115	+.2 to +.5	15% thick; gentle stall; good lift at altitude
23015	1.48	15°	−.008	.0102	0 to +.3	15% thick; sharp stall
63₂-215	1.41	14°	−.035	.010	−.1 to +.4	15% thick @ .35 chord; laminar flow—fair stall
64₂A215	1.35	15°	−.035	.0102	0 to +.4	15% thick @ .35 chord; laminar flow—gentle stall
65₂-215	1.41	16°	−.035	.0104	−.1 to +.3	15% thick @ .40 chord; laminar flow—good stall
65₂-415	1.44	16°	−.061	.0102	+.2 to +.6	15% thick @ .40 chord —laminar Cherokee section —gentle stall
66₂-215	1.41	16°	−.035	.010	−.1 to +.3	15% thick @ .45 chord —laminar—good stall
GA(W)-1	1.52	15.5°	−.09	.012	0 to +.8	17% thick @ .40 chord —gentle stall —new NASA section
GA(W)-2	1.62	16°	−.105	.011	0 to +.6	13% thick @ .40 chord —good stall—latest NASA section

* RN = Reynolds number = 9360 × V_{mph} × wing chord in feet (an empirical comparison value for flight speed and wing chord).

† α = angle of attack

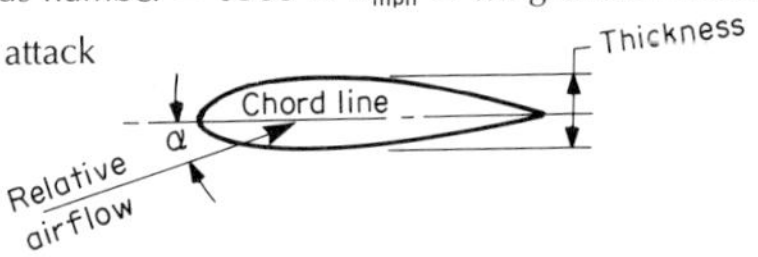

FIGURE 6-2
Airfoil comparison data (see Ref. 6.3)

lift values with no drag penalty at cruising speed, their moment coefficients are as much as 3 times those of the older and equally suitable 4415 and laminar-flow airfoils and over 11 times that of the 23015 section. In final analysis, the airfoil selected is usually a compromise between the structural considerations of maximum thickness and beam location; section drag at cruise; the lift coefficient the designer would like to have; and the tail load the airplane can carry without penalizing either structural weight or performance. For those readers interested in further airfoil study or data, Ref. 6.3 reports will be most helpful.

Wing flaps are the third means of increasing total lift and are capable of reducing stall and landing speeds beyond the limits possible with wing area increase or through selection of a different airfoil. As shown by Fig. 6-3, flaps come in many shapes and applications, but the type most commonly used on small aircraft is the slotted flap. Although the Fowler flap provides a greater increase in lift coefficient, as plotted on Fig. 6-4, the rather complex and heavy linkage necessary to extend this type of flap behind the wing trailing edge is really too expensive for the added gain. Actually, Fowler flap lift is realized by an increase in both lift coefficient and total wing area resulting from the addition of the extended flap planform to the basic wing area.

Slotted flaps are usually hinged at a fixed point below the flap lower surface, as may be seen on the Cherokee and other aircraft. The Cessna line uses an adaptation or combination of the slotted and Fowler types which is track-mounted but does not go as far aft as a true Fowler flap and, in fact, behaves like a slotted flap.

The comparative gain in lift coefficient ΔC_L possible with these various flap arrangements is given by Fig. 6-5, along with the drag coefficient increases and section moment coefficients for flaps lowered 45°. This is about the maximum

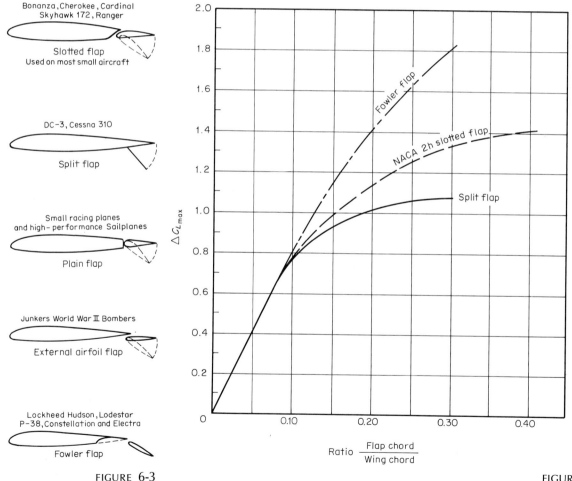

FIGURE 6-3
NACA 23012 airfoil with different types of flap

FIGURE 6-4
Comparison data, NACA 23012 airfoil section with various types of flaps—increase in ΔC_L vs. flap chord.

No.	Designation	Wing section	C_{Lmax}	$\alpha°$ at C_{Lmax}	C_{dmin} (flap down)	C_{dmin} (flap up)	C_{Mac}	NACA reference
1	Basic airfoil		1.4	17	. . .	0.010	0.07	Rept. 352
2	Simple flap on M6		2.1	16	0.07	0.011	0.22	Rept. 260
3	Split T.E. flap CY		2.1	14	0.19	0.011	0.26	T.N. 422
4	Simple flap on CY		2.2	13	0.15	0.011	0.30	Rept. 427
5	Hall flap on CY slot closed		2.2	13	0.17	0.011	0.29	T.N. 417
6	Zap flap CY		2.4	14	0.19	0.011	0.40	T.N. 422
7	Symmetrical T.E. airfoil on CY		2.5	12	0.06	0.013	0.34	T.N. 524
8	Fowler flap on CY		2.8	12	0.15	0.011	0.80	T.N. 419

FIGURE 6-5
Types of wing flaps—section coefficients compared

angle ever needed for small aircraft, since flap drag builds up rapidly beyond 40° deflection.

While you might have thought wing design was a simple process, quite apparently much compromise is involved—starting with takeoff speed and gross weight requirements. By using a balanced combination of wing area, airfoil lift characteristics, and a trailing edge flap capable of nearly doubling lift over the flapped portion of the wing, the designer can offer a landing speed under 60 mph, flaps down; this, of course, also means a lower takeoff speed.

As any owner's manual will indicate, practically all modern private aircraft equal or better this touchdown speed. Too much wing area means a reduction in cruising speed due to all that surface drag, so of course our modern airplane is equipped with a version of slotted flaps running from the fuselage out to the ailerons. This addition is necessary to improve short-field takeoff and landing control, reduce the stalling speed, and meet FAA maximum permissible stalling speed requirements.

CLIMBING OUT Once takeoff speed has been reached, a gentle pull on the wheel or stick takes the airplane out of ground effect and into climbing flight. But (1) what makes an airplane climb, and (2) why do some designs climb faster than others of the same weight and power? The answers are (1) excess horsepower and (2) less drag. The solutions, however, are not that simple.

Excess or surplus power beyond that required for level flight is converted into climb in accordance with the relationship

$$\text{Rate of climb} = \frac{\text{thp avail.} \times 33,000}{W} \text{ (in ft/min)}$$

where thp avail. = thrust horsepower available (excess horsepower × propeller efficiency)

W = airplane weight, lb

The problem faced by the designer during this phase of flight operation is to obtain maximum thrust horsepower in combination with minimum airplane drag. To show just what is happening, let us refer to the classic performance diagram of Fig. 6-6, considering only climb-related details at this time.

The upper line is the power-available curve representing the net thrust horsepower vs. airspeed left to propel the airplane after accessory and propeller efficiency power losses have been deducted from engine-rated horsepower. The lower line is the thrust horsepower required by the airplane to maintain level flight at different airspeeds between stall and maximum horizontal flight speed. These curves are usually developed for the gross weight condition at sea-level standard-day temperature and pressure.

At some speed point, or possibly for a narrow range of airspeeds, the gap between the power-required and power-available curves will have maximum value; this is the speed for maximum rate of climb. In Fig. 6-6 maximum climb

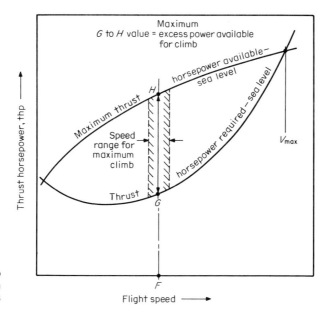

FIGURE 6-6
Aircraft performance diagram emphasizing climb parameters

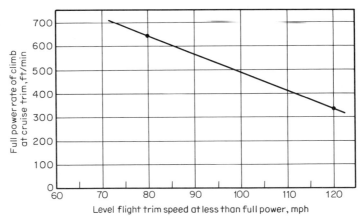

FIGURE 6-7
Full power climb from cruise
trim condition

occurs at speed *F*, as indicated by the maximum gap measured between points *G* and *H*, and extends over a narrow airspeed range of similar excess power. For an actual airplane there might be two or three power-required curves representing various flap extensions or other configuration changes; but regardless, at any given airspeed the distance between the two curves is the power available for climb at the particular airplane configuration represented by the power-required curve.

You can demonstrate this quite easily in your own airplane by following the procedure documented in Fig. 6-7. This is actually a magnified portion of the performance diagram based upon flight test data obtained with a 150-hp Cherokee.

When this airplane was trimmed at a level flight speed of 80 mph, use of full throttle power with no change in trim setting resulted in a steady climb rate of 640 ft/min. Similarly, when the airplane was trimmed at a level flight cruising speed of 120 mph, applying full throttle without touching the trim or elevator controls produced a steady climb rate of about 330 ft/min. The rate of climb was lower at the higher airspeed because a greater percentage of the total power was being used to obtain a faster cruising speed; as a result less power was available to provide climb. This may be clearly seen by the decreasing gap between the power-required and power-available curves of Fig. 6-6 as flight speed increases. But regardless, both rate of climb and therefore altitude are being controlled by the amount of excess power available.

Quite obviously the more we can raise the power-available curve and/or lower the power-required curve, the more thrust horsepower we will have for climb. Neither of these goals is easy to achieve.

The apparent simplicity of just adding horsepower to increase climb may be acceptable for specialized aircraft, but this approach is not completely ideal for production designs since empty weight, initial cost, and fuel consumption will all be increased as noted in the takeoff discussion.

Improved propeller efficiency, however, does represent a real gain at little or no weight penalty since thrust horsepower is a function of engine power and propeller efficiency. The efficiency of an installed propeller rarely exceeds 75 percent because of cowling blockage and similar losses, so an efficiency increase of 2.50 percent represents a power increase of 77.50/75 or a bit over 3 percent. This is a real gain as all the added power can be used to increase climb. Such increases in efficiency may be realized through improved blade design or refined cowling development.

Just as a constant speed propeller improves takeoff thrust, it will also increase climb rate by providing more efficient blade angle settings at climb speed. A fixed blade angle propeller set for maximum climb performance will unload and overspeed in cruising flight in similar manner to the fixed pitch takeoff prop previously discussed. So a constant speed propeller is desirable for maximum takeoff *and* climb performance without any sacrifice of cruising speed. Of course, some penalty is incurred since this type is both heavier and more expensive than wooden or forged aluminum alloy fixed pitch propellers. If maximum climb rate rather than cruising speed is the design condition, as would be the case for a towplane used as a sailplane tug, the fixed pitch propeller is acceptable and preferred; but such applications are in the minority and really represent specialized design requirements.

Another, and increasingly frequent, means of obtaining more power is by turbocharging or impeller-blowing the engine intake air. While these forms of supercharging will be described in greater detail in Chap. 11, the principle basically consists of compressing the induction air sufficiently to develop full rated power from sea level up to some cruising altitude such as 12,000 to 18,000 ft. Since basic induction system losses regularly steal 5 to 10 percent of rated power at sea level, supercharging provides added climb power from sea level right on up through altitude cruise. Of course the availability of full power at altitude is a major safety feature during operation from high altitude airports and lakes, to say nothing of clearing the Western Rockies and similar mountain ranges in Canada, South America, and Europe.

Outside of these three ways to increase the available climb power—namely, using a more powerful engine, developing a more efficient propeller installation, and turbocharging or blowing the induction air—there is little more a designer can do to increase thrust horsepower other than eliminate all accessory drives. This means that for a given powerplant installation, any increase in climb will have to be obtained by reducing the amount of thrust horsepower required.

INCREASING CLIMB   Since the required-power curve is based primarily upon airplane drag and to a lesser extent airplane weight, a reduction in either or both of these values will reduce the power necessary to maintain level flight and also increase the rate of climb for a given airspeed.

Reference to our climb formula on page 79 indicates that rate of climb

(ROC) will vary inversely with weight; i.e., the ROC increases in direct ratio to weight reduction. This means that cutting the climb weight in half would double ROC at a given speed and power setting. I think this clearly shows how overloading an airplane can lead to reduced climb, particularly on a hot day—and still worse, if the day is hot and the altitude high. Quite obviously one of the best ways to climb fast is to climb light, a fact we all know but sometimes forget to apply.

Given a fixed gross weight and powerplant installation, the most efficient and permanent method of increasing climb is through drag reduction. In addition to improving flight safety and operating economy, the attack upon drag reduction has the capability of being launched from many fronts.

The first, largest, and most effective drag reduction source is that of proper wing design. As wing area is increased at a given flight weight, the lift coefficient required will be reduced proportionately in accordance with the basic formula:

$$L = C_L \cdot \frac{\rho}{2} \cdot S_w \cdot (v)^2$$

where L = lift (airplane weight), lb

C_L = lift coefficient

ρ = mass density of air = 0.002378 at sea level

S_w = wing area, sq ft (including fuselage area as shown on Fig. 6-8)

V = airplane speed, ft/sec

As you will note, for given values of air density ρ, airplane weight L, and airspeed V, C_L will reduce in direct ratio to increased S_w. This reduced C_L value will in turn lower the wing *induced drag* C_{D_i}. Reduction of induced drag is important since this type of drag represents a considerable portion of the total airplane drag at climb speeds, acting in accordance with

$$C_{D_i} = \frac{(C_L)^2}{\pi \cdot AR}$$

where C_{D_i} = induced drag coefficient

C_L = wing lift coefficient required at the airplane weight and flight speed

π = 3.14

AR = wing aspect ratio = $\dfrac{(span)^2}{area}$

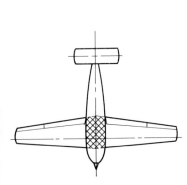

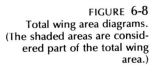

FIGURE 6-8
Total wing area diagrams. (The shaded areas are considered part of the total wing area.)

This formula tells us that cutting the lift coefficient in half will reduce the induced drag coefficient by 75 percent. However, a reduction of this magnitude

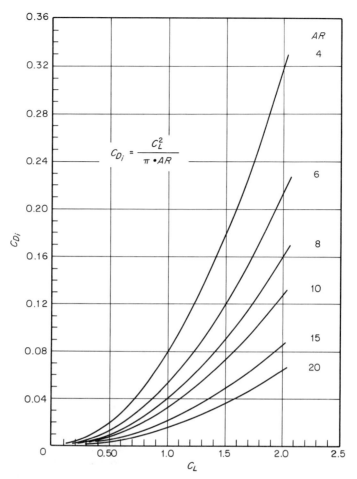

$$C_{D_i} = \frac{C_L^2}{\pi \cdot AR}$$

FIGURE **6-9**
Induced drag C_{D_i} variation
with lift C_L and aspect ratio
AR (*AR* correction factor = 1.0)

is not practical because the required wing area increase would be prohibitive, as previously noted. Although some increase in wing area should be used to reduce induced drag to an acceptable level, the added wing surface area increases skin friction drag and airframe weight—both of which tend to lower climb rate. So again the designer is faced with a trade-off compromise: this time between desired wing area vs. resulting structural weight and skin friction drag increases.

Referring again to the induced drag formula, we see that increased aspect ratio also acts to lower C_{D_i}, although not to the same degree as reduced C_L. Since increased aspect ratio has a favorable effect upon cruising flight drag, it is a desirable design feature, provided any related structural weight increases can be kept under control.

The effect of aspect ratio upon induced drag for different lift coefficients is presented in Fig. 6-9, which is a graphic presentation of the induced drag formula. Note that at a cruising C_L of 0.50 the induced drag coefficient is 0.02 at an aspect ratio of 4, but only 0.004 at *AR* of 20. Five times the aspect ratio results in one-fifth the induced drag. Since an aspect ratio of 20 is really too high to be practical for general aviation use from both structural weight and ground handling considerations, some compromise is necessary—possibly finalizing the design at an aspect ratio value of 8 with accompanying induced drag coefficient of 0.01.

It is also apparent that high aspect ratio is not important to aircraft designed primarily for high-speed operation at low values of C_L. At a speed requiring a C_L of only 0.25, induced drag changes vary slightly whether the wing aspect ratio is 4, 8, or 20. However, general aviation aircraft, even those designed for high-speed cruising performance at low C_L values, have to climb to operating altitudes at speeds below cruise. At the comparatively high lift coefficients required during climb, increasing the wing aspect ratio will lower the induced drag, which in turn releases horsepower to increase the climb rate.

Therefore, for a given flight weight, wing area, powerplant installation, and similar general configuration it may be stated that: *The airplane with the greatest aspect ratio will have the highest rate of climb.* Again some trade-off is necessary since wing structural weight increases with aspect ratio. This happens because the wing root and inboard section bending moments increase as the center of lift moves outboard with increased wingspan. If the wing weight is allowed to get too high, the loss in ROC due to weight increase could exceed any drag reduction realized through increased aspect ratio.

The reduction of interference drag is another important improvement that favorably affects both climb and cruise performance. This type of drag is caused by the breakdown of airflow over external structural and accessory items such as landing gear struts and wheels, skis, wing and tail brace struts, wires, exposed bolt heads and nuts, drain fittings, radio antennas, and similar appendages. In fact, the reduction of interference drag probably represents your maximum opportunity for performance improvement.

In the landing gear area alone, design comparison studies have shown that adding wheel fairings to tricycle gear aircraft will improve climb performance midway to that of fully retracted gear. Referring to Fig. 6-10, if a design climbs 750 ft/min with unfaired tricycle gear lowered and 1000 ft/min with the gear retracted, the fixed gear climb performance will be increased to about 900 ft/min by adding wheel fairings to all three legs. Faired gear can reduce the performance attractiveness of retractable landing gear to such a level that the complication hardly seems justified for most general aviation designs—so don't leave your wheel fairings behind, particularly in hot weather when climb becomes increasingly critical.

Further drag reduction is possible by sealing cowl and cabin leaks, not leaving the cowl flaps fully open on cold days, keeping the entire airplane polished and free of mud and bugs, streamlining or using buried antennas, keeping the wing and tail surface leading edges free of dents, and filleting wing and

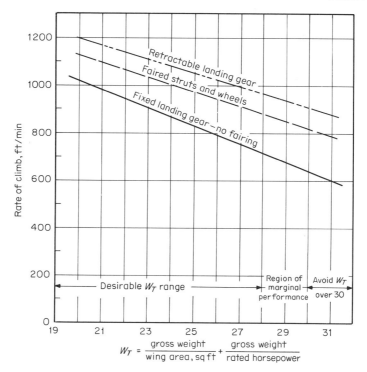

FIGURE 6-10
Landplane (tricycle landing
gear) climb rate vs. landing
gear fairing and W_T

canopy intersections. You can take care of many of these items yourself; each one will slightly reduce total drag and so contribute a bit toward improved climb and cruise performance.

In addition to the effects of weight and drag, climb is also affected by density altitude—that annoying combination of temperature and pressure altitude. Figure 6-11 (based upon FAA CAM 8, Fig. 9-2) shows that ROC decreases with increases in either pressure altitude or ambient temperature. This deterioration of performance occurs because engine power and propeller thrust decrease with increasing temperature and/or altitude—both of which reduce air density. As with takeoff distance, consult a Denalt computer for ROC when temperature and altitude conditions vary from the familiar pattern.

Although the FAA requirements of Ref. 6.4 will accept climb rates as low as 300 ft/min depending upon stalling speed in the takeoff configuration, modern aircraft climb in excess of 600 ft/min at gross weight and sea-level standard conditions. For safe operation, this should be the minimum acceptable ROC for aircraft design.

Flaps may or may not increase ROC depending upon the type of flap, flap extension, wing area, configuration drag, and climb speed. Certainly the drag producing split flap will increase neither climb rate nor angle of climb, while most other types can at least increase the climb angle by permitting slower climb speeds at about the same ROC. In this regard, the slotted flap and ver-

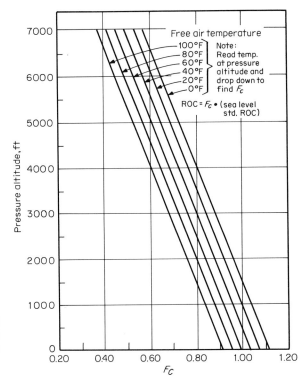

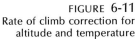

FIGURE **6-11**
Rate of climb correction for
altitude and temperature

sions of the Fowler wing area increasing flap are the best types for short-field climb-out over local obstacles and hillsides, as previously reviewed under takeoff design.

While I hesitate to enter the arena, some discussion regarding the highly touted "turned down" wing tips is probably warranted at this point. Personally conducted flight test investigation of such tip additions to the Teal amphibian showed slightly improved lateral stability in climb and approach, but no measurable increase in climb or reduction in power "off" descent rate. However, on a high aspect ratio competition sailplane, sculptured wing tips did reduce the stalling speed while also improving lateral stability to a slight degree. On balance, for production private aircraft it is doubtful whether the gain in climb performance exceeds that realized from the wing area these larger tips add to the basic planform. Quite likely the Mooney "flat rib" wing tip is as effective as any other and much cheaper as well—at any rate, it certainly works well for Mooney aircraft and many homebuilders.

So much for the many factors affecting ROC. There is one other characteristic that will affect pilot acceptance of an airplane during climb, however, and that is *dynamic stability*. As you will recall we have previously discussed *static stability,* which may be ideally demonstrated in flight by stick forces that increase linearly as the airplane is displaced from a set of trimmed conditions. This effect is quite similar to a moored boat righting itself from local waves or from someone moving about on the deck. If this same boat when underway recovers equally well from pitch or roll displacement through a series of decreasing oscillations, it is considered to be dynamically stable; but if the boat tends to roll over or pitch under at speed in a seaway instead of recovering to its original upright trim condition, it is dynamically unstable.

In similar fashion, if an airplane recovers from displacement in pitch, roll, or yaw caused by local turbulence, overcontrol, or normal flight maneuvers, it is considered dynamically stable. If overcontrol or displacement about any of the three axes results in a continued or increasingly divergent oscillation, the airplane is dynamically unstable. This condition is very undesirable for any private airplane since it will require constant pilot attention to maintain a heading, desired altitude, or trim condition.

Nothing can be more annoying than an airplane that constantly oscillates about one or more axes during climb, particularly in turbulent air conditions. For this reason current FAA design requirements place heavy emphasis upon stability about all three axes—pitch, roll, and yaw—and throughout all phases of flight. This is why your modern production FAA certified airplane does not exhibit unstable tendencies in climb so common to early aircraft of all types.

At some altitude, all thrust horsepower available will be required to keep the airplane in level flight at a speed just above indicated stalling speed. The altitude at which this occurs is known as the *absolute ceiling;* the airplane cannot climb higher for a given takeoff weight. As another standard of performance, the altitude at which ROC has decreased to 100 ft/min is known as the *service ceiling;* this will be lower than the absolute ceiling and represents an altitude at which some maneuvering is possible provided all controls are well coordinated.

As a final note, sea-level climb rate and service ceiling are important design points for single-engine aircraft, while the single-engine ceiling represents a major design consideration for twin-engine aircraft. For all normal aircraft, sea-level climb, service ceiling, and single-engine ceiling are improved by installing either constant speed propellers or turbocharged engines or both. As described in Chap. 11, altitude performance is greatly improved when turbocharging and constant speed propeller operation are combined.

In summary, since maximum ROC occurs at the speed for airplane maximum lift to drag ratio L/D, the lower the total drag the greater the ROC. In addition to this basic drag consideration, ROC is also a function of available power, airplane weight, and density altitude. As we have seen, satisfying any one of these factors affects all the others, requiring considerable design trade-off study to obtain acceptable climb performance within practical horsepower, wing area, useful load, and operating altitude limitations.

REFERENCES   6.1   FAR 23.49(b).

6.2   FAR 23.51(b) (2).

6.3   (a) Jacobs and Abbott, *Airfoil Section Data Obtained in the NACA Variable-Density Tunnel.* NACA Technical Report No. 669, 1939.
      (b) Abbott and von Doenhoff, *Theory of Wing Sections—Including a Summary of Airfoil Data.* Dover Publications, Inc., New York, 1959.

6.4   FAR 23.65(b).

7 CRUISING WITH STABILITY

After climbing to the desired altitude, we begin cruising to our destination—and this is the phase of flight for which most private aircraft are purchased. Whether used for business or personal convenience, modern aircraft offer a great time-saving advantage over the automobile because they can fly between two points in an essentially straight line. The faster the speed along this line for a given powerplant size, the more economical the airplane. These are rather obvious facts, but the ways to attain them are not so obvious.

WHAT PRICE CRUISING SPEED? Increasing the size of the engine will raise the cruising speed, but, as with take-off and climb, at the sacrifice of useful load and fuel economy as well as increased first cost. And the speed increase with power change will not be all that great, since for the same altitude speed varies in direct ratio to the cube root of the power change, thus

$$V_2 = V_1 \sqrt[3]{\frac{(hp)_2}{(hp)_1}}$$

where V_1 and $(hp)_1$ = the original speed and horsepower

V_2 and $(hp)_2$ = the new speed and increased horsepower

If we cruise at 130 mph with 150 hp, this formula tells us the cruising speed will increase to 143 mph when a 200-hp engine is installed, offering a gain of 10 percent in cruising speed for an increase of 33 percent in power and fuel consumption. This is not a particularly exciting trade-off, but it is typical.

For example, the current production 180-hp Beechcraft Sundowner cruises at 136 mph with fixed tricycle gear, while the 200-hp Beechcraft Sierra of identical wing area cruises at 151 mph with retractable gear and a slightly higher gross weight. These differences are important design considerations, however, because increasing airplane weight with the same wing area means the lift coefficient and wing drag will also increase slightly. Since our formula tells us the 200 hp Sierra should cruise at 141 mph with fixed gear vs. the published cruise of 151 mph with retractable gear, we see that by retracting the Sierra landing gear Beechcraft engineers were able to reduce cruising drag sufficiently to carry the increased gross weight with some gain in speed—but at the price of increased empty weight, initial cost, fuel consumption, and maintenance. In view of this, what practical alternatives do we have available to increase cruising speed and fuel economy?

By way of defining this speed problem let us again refer to the aircraft performance diagram, now adding the cruising power-available curves to obtain Fig. 7-1. This picture graphically shows what determines (or limits) flight speed, while also suggesting ways that speed may be increased for more economical operation.

The most effective approach to increased efficiency is through reduction of airplane drag, which in turn will reduce the thrust horsepower required at all flight speeds. As you can see, at 65 percent power our airplane is limited to cruising speed C reached at point D, which is formed by the intersection of the power-available and the power-required curves. This intersection determines level flight speed for a given altitude and power selection; if there is no more power available at a given throttle setting, the airplane just cannot fly any faster at that altitude.

Assuming these curves are given for sea-level operation which is the usual procedure, it will be possible to decrease cruising drag somewhat by climbing above sea level to regions of reduced air density. Up to the altitude requiring full throttle operation to obtain the power desired, engine cruise power may be maintained at a constant percentage of sea-level power by steadily advancing

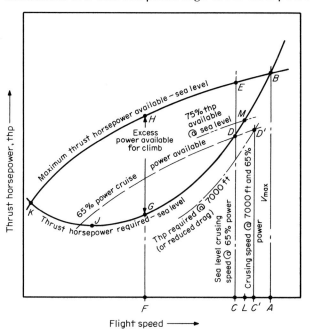

FIGURE 7-1
Aircraft performance diagram
showing speed limits

the throttle. Since airplane drag decreases with air density, the thrust horse-power required to maintain level flight will decrease with altitude. This means that as long as the desired percentage of sea-level cruising power can be main-tained, cruising speed at that power will increase with altitude up to the flight level requiring full throttle operation.

This explains why an unsupercharged airplane can cruise at increasing air-speed with increasing altitude up to some power-limited speed such as C', de-termined by point D' intersection of Fig. 7-1. For modern private aircraft this point is reached between 6500 to 8000 ft depending upon the powerplant in-duction system. So if you have a long flight ahead, this is why you should cruise at altitude—weather permitting. For some propeller-driven aircraft, full throttle cruising speed at altitude will exceed full throttle V_{max} at sea level.

All of this indicates the obvious need to reduce drag in order to increase cruising speed and airplane efficiency. As one approach to this goal, the designer can reduce frontal area, which usually means a loss of elbow and head room or increased cabin intimacy à la early Mooney aircraft (which are now improved inside with the same width outside). Sailplanes carry this con-cept to the nth degree by reclining the pilot to such an extent it is a wonder we don't experience accidents caused by dozing pilots. But since the sailplane power source is gravity, the less drag, the less descent required; and by ever re-ducing sailplane frontal area and interference drag the design glide ratio stead-ily advances toward the magic value of 40 to 1.

As another approach, airplane cruising drag can be reduced by decreasing wing area, but overdoing this will reduce climb rate and increase takeoff and landing speed as discussed in Chap. 6. The necessary design trade-off usually requires the use of flaps to obtain landing speeds acceptable to the FAA, plus more wing area than desirable for maximum cruise in order to provide min-imum acceptable climb rates.

Although a specialized design type, the early Cleveland Air Racers illustrated this point very well. Their wing area was reduced to such a degree that extremely long takeoff runs and very shallow climb angles were accepted in order to obtain the speeds required from the powerplants available. In fact some of these designs were barely able to get airborne with their design wing area/powerplant combinations.

Thinner wing sections, increased aspect ratio, improved engine-cooling air-flow, faired canopy and cowling assemblies, wheel fairings or retractable land-ing gear, buried antennas, sealed control surfaces, and many similar items all contribute to drag reduction. These possibilities are explained at greater length in Chap. 10 for those further interested in this subject. For our purposes at this time it will be sufficient to note that a reduction in airplane drag will lower the required thrust horsepower. For comparison purposes let us consider a reduced-drag power-required curve lying below the sea-level curve values shown on Fig. 7-1. This means that through reduced airplane drag an intersec-tion similar to D' could be realized at sea-level speed C', with higher cruising speeds available at somewhat greater altitudes. Such an improvement represents a decided gain in aircraft performance and efficiency.

Following this cruising speed discussion a bit further, if the throttle is advanced to 75 percent power at sea level we obtain a higher cruising speed L resulting from the intersection at point M of the 75 percent thrust horsepower available and the basic sea-level power-required curves. In similar manner power can be increased right up to point B, resulting in sea-level maximum speed A. Again, this speed would be expected to increase with altitude up to some relatively low flight level depending upon the powerplant design and airplane configuration.

Since drag reduction offers improved operating efficiency throughout the entire flight profile from takeoff to landing, it is important that all airplane surfaces be kept clean and polished and all air leaks sealed to the extent possible. Drag reduction work is time well spent to obtain a safer and more efficient airplane.

FLIGHT STABILITY

Aircraft certified within the past 10 years must satisfy very demanding stability requirements. For example, modern aircraft may not be demonstrated to meet certain directional stability criteria by use of the rudder; the airplane design must be such that specified directional stability limits can be met by the airplane itself without pilot assistance. In addition, if a rudder trim system is used, it must be adequate to fly the airplane if the primary rudder controls should fail—a very remote possibility at best and really an equally dubious requirement for small aircraft.

Longitudinal stability (about the pitch axis) must be so positive that little trim change or stick force increase is developed when executing a missed approach or balked landing with the airplane fully trimmed in the landing configuration. Minimum trim change with application of flaps or power, plus the requirement that an airplane can be flown with the longitudinal trim system (elevator tab or adjustable stabilizer) are other major longitudinal stability criteria; but there are many more.

To round out the stability picture, amendment 14 to FAR 23 added a requirement for minimum acceptable rates of roll during takeoff as well as in the landing configuration; so now all new private aircraft must meet rigorous minimum stability requirements about all three principal flight axes. In fact, once trimmed an airplane really must be able to fly itself in smooth air. The balance of this chapter will consider how this level of flight stability may be realized as well as why it is desirable and important. Once again, the designer is faced with a very interrelated set of configuration problems.

LONGITUDINAL STABILITY AND TAIL POWER

Lateral and directional stability effects tend to act together. For example, a rather rapid application of rudder will raise the outside wing in a turn because of increased lift on the outside panel. In this case the rudder and fin produced turning lift (directional control and stability) which rotated the airplane about a vertical axis and so accelerated airflow over the wing panel on the outside of the turn. This accelerated flow produced more lift on the outside panel than on the inner one and this made the airplane bank into the turn (lateral control and stability) by rotating about the fore and aft axis. This is typical of the lateral directional stability relationship. However, longitudinal, or pitch stability, acts quite independently of either of these axes and is the most important of all three.

The primary requirement for longitudinal stability is adequate horizontal tail power. *Horizontal tail power* is a combination of horizontal *tail area* and the *tail arm,* the distance from the airplane center of gravity to the horizontal tail center of lift. The greater the tail arm the smaller the horizontal tail can be, explaining why sailplanes have long tail arms and small tail surfaces to reduce both tail surface drag and control surface displacement drag.

These same options are available to the powered-aircraft designer but, since powered-aircraft horizontal tail loads are greater than those for sailplanes, some structural weight penalty is incurred when an aft fuselage becomes overly long. On the other hand when the tail arm is too short, any airplane will tend to become marginally stable longitudinally, and finally unstable in pitch as the center of gravity moves aft.

FAA flight test requirements include demonstration of acceptable longitudinal stability at a gross weight center of gravity located further aft than the most rearward center-of-gravity limit to be finally approved for certification. Since this aft point is often determined by pilot opinion unless some dangerous spinning or recovery tendency has been exhibited during flight test, a lively discussion frequently develops between the FAA and the manufacturer's personnel at this point in the flight demonstration program. However, rest assured that if you fly within the aft center-of-gravity limits for your airplane, you will always have positive stability and should never experience a reversal of stick force under any flight condition. But the degree or level of positive longitudinal stability will steadily decrease as the center of gravity moves aft because the tail arm is decreasing.

For the benefit of those interested in aircraft design and as a good first approximation, the one-third chord point of the horizontal tail should be located 2.75 to 3 times the wing mean aerodynamic chord (MAC) distance from the wing quarter chord point (or airplane center of gravity).

As previously noted in Chap. 1, the longer the tail arm the more stable an airplane will feel and actually be in flight. The rather simple reason being: the longer the tail arm, the more damping action provided by the horizontal tail. To meet FAA design and flight test longitudinal stability requirements while developing an airplane with acceptably low elevator control forces under all flight conditions, the designer has to consider and optimize: tail arm, horizontal tail area and lift coefficient, surface drag, structural weight, and the type of horizontal tail design to be used.

TAIL SELECTION   Tail surface design should be covered at some point in the text, and this stability discussion seems a logical spot since the horizontal tail is the principal determinant of longitudinal stability.

An airfoil suitable for tail design will usually be of symmetrical section, and preferably no thicker than 12 percent of the chord length. Although the tendency would naturally be toward selection of a thicker section to reduce bending stresses, the airfoil section *lift slope* falls off rapidly above 12 percent thickness while, of course, the section drag increases. The lift slope of an airfoil is based upon the straight portion of the lift curve and is the ratio of lift coefficient increase per degree change in angle of attack. The higher the lift slope value the greater the tail lift and damping action, which in turn means increased longitudinal stability and control response.

As a matter of interest, for suitable tail surface airfoils the reports of Ref. 6.3 show the following decreases in lift slope with increased section thickness:

NACA section	Lift slope value
0006	0.1125
0009	0.1094
0012	0.1094
0015	0.0750

For those readers interested in the design of horizontal tails providing an acceptable level of longitudinal stability, the following formula permits simple calculation of the minimum surface required in relation to wing area and tail arm:

$$P_{HT} = \frac{D_H}{MAC} \times \frac{S_{HT}}{S_w} \geqslant 0.55$$

where P_{HT} = horizontal tail power, an empirical value

D_H = distance from center of gravity to 0.30 chord horizontal tail, ft (horizontal tail arm)

MAC = wing mean aerodynamic chord (or average wing chord), ft

S_{HT} = total horizontal tail area, sq ft

S_w = total wing area, sq ft

According to NACA test results of Ref. 7.1, the all-movable and trimming stabilizer configurations require less area than a fixed stabilizer, so the above P_{HT} value could be reduced about 10 percent for such designs.

Similar to the determination of an adequate horizontal tail, vertical surface area should equal or exceed the value

$$P_{VT} = \frac{D_v}{MAC} \times \frac{S_{VT}}{S_w} \geqslant 0.30$$

where P_{VT} = vertical tail power, an empirical value

D_v = distance from center of gravity to 0.30 chord vertical tail, ft (vertical tail arm)

MAC = wing mean aerodynamic chord (or average wing chord), ft

S_{VT} = total vertical tail area, sq ft

S_w = total wing area, sq ft

While the planform design of the horizontal or vertical tail is frequently as much a matter of artistic effort or structural simplicity as aerodynamic consideration, tail aspect ratio should be as high as weight allowances will permit—which really means the aspect ratio should be as high as structural loading permits. This is generally around 3 to 4 for horizontal tails and 2 to 3 for vertical surfaces.

In summary, to be effective aerodynamically and still remain within structural weight limitations, all modern tail surfaces are of fairly thin section and relatively low aspect ratio.

Horizontal tail position on the fuselage or fin is not so typically standardized or well defined. We have low tails, mid tails, T tails, upright V tails, and inverted V tails—all with different effects upon damping, control response, control system design, structural weight, and, possibly most important of all, spin recovery effectiveness.

To go back a few generations in design, the designers of the first airplanes not only followed a random approach for vertical location and planform of the horizontal tail, but also frequently placed the tail forward, or aft, or at both locations. Although they soon became leaders in elevator surface and control design, I'm sure the Wright brothers' long periods of tethered glider flight were primarily elevator system design, development, and training flights; they were learning the response of the airplane to elevator trim changes. No one had ever previously documented this phenomenon, nor was there anyone around to teach the required technique. So the Wrights located the elevator ahead of the airplane where it could be closely observed in operation, and learned to fly. But many builders immediately adopted this tail location because the Wrights were so extremely successful for their time.

Fortunately, as flyable aircraft designs increased in number, Bleriot, Antoinette, REP, and many others realized why an arrow has feathers at the aft end. By 1911–1912 most designers had located the horizontal tail behind the wings where it would be most effective at low airspeeds. From the structural design standpoint the aft and low horizontal tail position was and still is the lightest arrangement since it is the easiest to attach to the fuselage. The earliest aircraft simply tied the fixed portion of the horizontal tail or tails to the fuselage booms which provided a convenient location for a light surface. Figure 7-2 shows this type of mounting on a 1911 Voisin apparently designed with one of the first T tails. Early Curtiss aircraft had similar rather fragile tail mountings as indicated by Fig. 1-2.

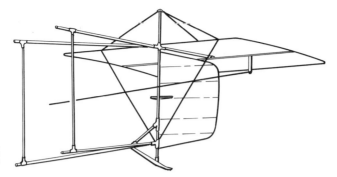

FIGURE 7-2
1911 Voisin tail assembly

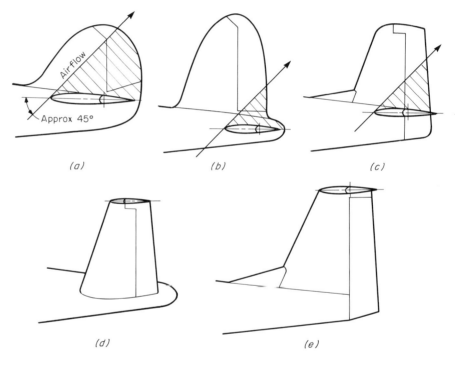

FIGURE 7-3
The effect of horizontal location upon rudder response during spin recovery. (a) An early type of rudder design subject to blanketing by the horizontal tail. (b) Horizontal tail location for improved rudder effectiveness during spin recovery. (c) Rudder extended below the horizontal tail for increased effectiveness during spin recovery. (d) Bonanza type V-tail design provides fully effective tail surfaces during spin recovery (inverted V tail is similar). (e) The T-tail configuration has the fin and rudder completely effective with no blanketing.

When higher speeds were realized during World War I, the need to reduce turbulence generated by exposed struts and flying wires as well as from slipstream passing over unfaired tail booms became quite obvious. As a result, fuselages were enclosed around the tail surfaces and so the low to mid position of the horizontal tail developed as the standard location for many years.

When the rudder was not carried down below the horizontal tail, frequently the case with early designs, some aircraft experienced extreme difficulty in recovering from either intentional or inadvertent spins, a condition caused by horizontal tail blanketing of the rudder, as shown in Fig. 7-3a. This type of blanketing not only prevented rudder reversal from stopping spin rotation, but also frequently contributed to rudder lock in the direction of spin rotation—with fatal results. Subsequent spin studies produced the recommended horizontal tail locations of Figs. 7-3*b* and 7-3*c* as well as the extended rudder shown in *c*.

As a further refinement, right after World War II Beechcraft developed the distinctive V tail of the Bonanza series to provide increased horizontal and vertical tail effectiveness throughout the entire flight spectrum. Some drag and cost reduction were also anticipated as the result of having only two surface assemblies instead of the customary three. However, like some modern corrective medicines the V-tail configuration has some related side effects that are not wholly desirable, as will be noted a bit further on.

At present, all these horizontal tail locations are gradually being displaced by the T-tail arrangement of Figs. 7-3*e* and 7-4. While some of the earliest aircraft may have carried a T tail more as a matter of convenience than aerodynamic refinement, Focke-Wulf of Germany developed the first really engineered T-tail design for their late-World War II model Ta-183 jet fighter. In fact their initial design went all the way with a swept horizontal tail mounted on top of a swept fin—and with horizontal tail dihedral as well to provide the ultimate in T-tail design.

The T tail has been accepted rather slowly for private aircraft because it is generally heavier than a conventional tail assembly due to combined horizontal tail and fin bending loads which must be carried by the fin and fuselage

FIGURE 7-4
T tails arrive at Beechcraft.
Beech Aircraft Corporation Photo

structures. The control system is also more complex and somewhat heavier than for a lower horizontal tail location. There is some gain in directional stability, however, since the end plate effect of the horizontal tail increases fin and rudder effectiveness, while of course the entire vertical tail acts to assist in spin recovery. With careful design the increased vertical tail effectiveness can be used to reduce vertical tail area and weight, so some reduction in penalty is possible. For readers interested in T-tail design procedures, a few useful reports are carried in Ref. 7.2.

The action of the various tail arrangements in maneuvering flight and gust conditions seems to be an item of interest to everyone. Briefly, and with reference to Fig. 7-5a, the conventional low horizontal tail is affected by vertical gust conditions, while only the vertical tail responds to horizontal gusts. The elevator and rudder act quite independently to provide pitch and yaw control.

There will be some increase in surface pressure acting over the horizontal tail when the vertical tail is loaded by rudder displacement or horizontal gust action, as shown by the load arrows of Fig. 7-5a. In addition, downward gusts will increase the pressure field acting on both the horizontal and the vertical tails as indicated; note that such loadings generally act symmetrically on each side of the vertical and horizontal surfaces.

It is the horizontal gusts or maneuver-induced airflows which normally develop tail surface unsymmetrical loadings, but these values are usually quite small for the conventional horizontal tail position. For private aircraft, FAA design specifications only require a maximum differential of about 30 percent in loading between opposite sides of the horizontal tail, including gust and sideslip conditions (Ref. 7.3). In this design condition one side of the horizontal tail is loaded to 70 percent maximum design load while the opposite side is loaded at 100 percent.

Of course, unsymmetrical tail loading also places a twist, called torsion, on the fuselage aft section. Due to maneuvering or gust action, fuselage torsional loading is frequently further increased by the addition of vertical surface loads—and all these loadings acting together may require a very rugged aft fuselage structure as discussed under Chap. 12, Airframe Design.

The action of the V tails is considerably more complex than that of conventional surfaces. As you can see from Fig. 7-5b, the well-known Bonanza upright V tail (Fig. 1-5) acts to simultaneously yaw and pitch the airplane when encountering either vertical or horizontal gust action. This is why the Bonanza seems to yaw with pitch trim changes in turbulent air and vice versa, a condition similar to *hunting*, which is a term applied when an airplane oscillates in flight between two trim points. This characteristic can be nausea producing over prolonged periods in rough air, particularly for anyone who flies infrequently. Could this be the reason more conventional tail surfaces are used on the later Debonair and subsequent Beechcraft designs?

The inverted V tail of Figs. 7-5c and 7-6 exhibits the same interaction tendencies, but it has a slight design advantage over the upright V tail in that ver-

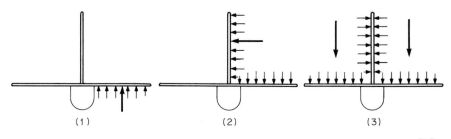

(1) (2) (3)

FIGURE *7-5a*

Loading conditions for the conventional tail surface configuration. Note that airflow may act on either side of the surface and that symmetrical loading of the horizontal tail may occur in an up or down loading direction. (1) Vertical gust or maneuver airflow acting on one side produces unsymmetrical loading on horizontal tail. (2) Horizontal gust or maneuver airflow loads vertical tail and increases pressure over one side of horizontal tail as well, producing unsymmetrical loading of horizontal tail. (3) Vertical down gust or maneuvering airflow acting on both sides increases the horizontal tail load. Balanced loads on the vertical tail result in symmetrical loading of horizontal and vertical tail surfaces.

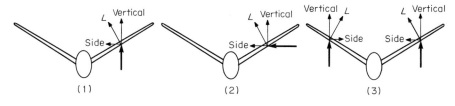

(1) (2) (3)

FIGURE *7-5b*

Gust and maneuver loading conditions for the upright V-tail configuration. Airflow may act on either side and in an up or down direction. (1) Lift component of vertical airflow produces both vertical and side loadings, moving airplane in pitch and yaw. (2) Lift component of horizontal airflow produces both vertical and side loadings, moving airplane in pitch and yaw. (3) Lift components of symmetrical vertical loading (up or down) act to move the airplane in pitch, but the side components cancel one another as well as any tendency to yaw.

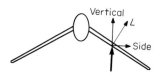

FIGURE *7-5c*

Inverted V-tail configuration. Lift component of vertical airflow produces both vertical and side loadings, moving the airplane in both pitch and yaw. The side force, however, acts in a direction opposite that of the upright V tail. Loadings are otherwise similar to those of Figure 7-5*b*.

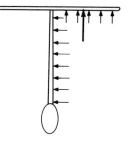

FIGURE *7-5d*

T-tail configuration under vertical airflow loading. Note pressure increase on vertical tail as well. Control surface deflections also increase loadings similar to gust action; otherwise, load conditions are similar to those for the conventional tail of Figure 7-5*a*.

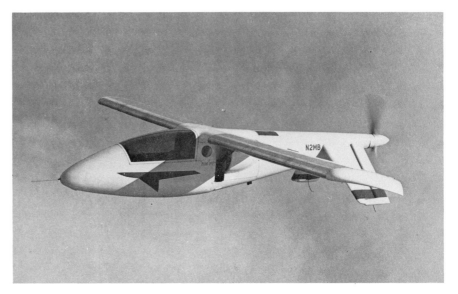

FIGURE 7-6
Mini Imp homebuilt with
inverted V tail
Taylor Photo

tical tail area located below the center of side area (approximately at the center of gravity) will tend to roll an inverted V-tail airplane into a banked turn. This means any tendency to be lax on rudder action is somewhat self-correcting on aircraft having inverted vertical tail surfaces. Since nothing is free in the world of aircraft design, the inverted tail presents a basic configuration design problem of providing adequate tail clearance for full stall landings. An overly long and heavy landing gear is the usual solution to this requirement, as well as being a reason the inverted V-tail arrangement is not popular.

Other objections to V tails in general are found in their more complex and somewhat heavier control systems, plus the heavier tail structure necessary to support combined horizontal and vertical surface loadings. In summary, the V tail is an unconventional arrangement which is generally as heavy as the conventional design it would replace, and the stability characteristics are somewhat less desirable—particularly in rough air, just when a high level of stability becomes increasingly important.

Our final configuration analysis covers the T tail, which is also subject to load interaction to some degree. Figure 7-5*d* shows how vertical or horizontal gusts and control-surface deflections tend to increase the air pressure acting over T-tail surfaces. Fortunately, these pressure changes are noted more for increasing structural loading rather than producing any adverse stability effects, since a T tail behaves like conventional surfaces.

The interaction of airloads with the T-tail configuration has been noted in a recent amendment to FAR 23 requiring combined application of vertical and horizontal tail surface loadings (Ref. 7.4). So although the T tail offers stability advantages without any penalty in landing gear design, a penalty is incurred through increased weight of the fin and aft fuselage structures. This penalty may be somewhat reduced by the increased fin area effectiveness previously noted, but the T-tail structure will still be somewhat heavier than a conventional tail arrangement. In addition, as for the V-tail design both the elevator control and longitudinal trim systems tend to become somewhat more complex and a bit heavier.

Regardless of these negative trade-offs, the T tail is gaining in popularity because of the improved stability offered for special design applications such as the minimum trim changes with power for the Teal amphibian (Fig. 4-13), and, more importantly, because of the excellent spin recovery characteristics offered by a completely unblanketed vertical tail.

Since rapid spin recovery is essential for reduction of low-altitude fatalities, designers are paying more attention to this most critical flight condition. Related test results and criteria for desirable spin recovery characteristics are given in Ref. 7.5. These investigations indicate we can expect to see more T-tail aircraft in the future. In the meanwhile, since there is little we can do to correct the stall/spin characteristics of existing aircraft, I'm sure that unfavorable accident statistics would continually decrease if everyone spent some time obtaining qualified flight instruction in spin recovery procedures.

OTHER STABILITY FACTORS

Although the entire subject of flight stability has filled many books, the primary requirements of center-of-gravity location, tail arm, and adequate tail surface design have just been covered in some depth. In addition, some other factors affecting cruising flight stability and handling characteristics should be briefly reviewed for background information. These are dihedral, adverse yaw, thrust line location, and the longitudinal trim system.

Although dihedral and adverse yaw have been discussed in Chap. 5, it is important to note that in turbulent air too much or too little dihedral will require constant aileron attention to correct lateral trim. If the aileron design develops any amount of adverse yaw, constant rudder correction will also be necessary in addition to all the other chores facing a pilot in rough air. This is a most annoying situation and an increasingly dangerous one to contend with when nearing minimum descent altitude levels during IFR operation. Dihedral is also tied to vertical tail area in developing adequate lateral stability. If the combination is not within certain limits an airplane can develop a steady rolling and yawing condition referred to as *Dutch roll*. While there are many explanations for this undesirable flight pattern, the airplane actually goes through a slow oscillation of 5 to 15° roll each side of the flight axis combined at the same time with about 5 to 10° yaw each side of the flight path. This behavior is similar to a runaway two-axis autopilot and, as can be imagined, would be most disturbing if experienced during flight—to say nothing of really blind IFR conditions.

FAA flight test procedures ensure that Dutch roll is not present in recently certified aircraft, but it can rear up in some pre-World War II designs when flying in turbulence. This stability problem is usually corrected by reducing the dihedral angle and/or increasing the vertical fin area. Military aircraft frequently resolve this condition by complex autopilot and feedback control system design. Fortunately that expensive approach is not necessary at private aircraft speeds. Instead, our airplanes should have good ailerons causing minimum adverse yaw, adequate vertical fin area, and not too much dihedral.

Minimum trim change with power is important during cruise since less pilot attention will be required to hold altitude or a desired heading. In view of this, the location of the thrust line relative to the center of gravity becomes another important design consideration; when the thrust axis passes far from the center of gravity, trim changes may accompany every change in power setting. Although only changes in pitch attitude with power will normally be developed with single-engine aircraft, either or both pitch and yaw trim variations should be expected from power changes on twin-engine designs where the thrust lines may be outboard from and above or below the center of gravity. Since the basics of this subject have been treated in Chap. 2 and Figs. 2-4 plus 3-4, only a special stability case will be mentioned at this point.

It is quite evident that, if any design ever did, the Teal amphibian of Fig. 4-13 really violates the principle of having the thrust line pass near the center of gravity. Yet because of the T-tail design and horizontal tail location, changes in slipstream velocity with power result in tail loadings almost exactly offsetting thrust moments about the center of gravity. So the T tail is a definite asset to an aircraft configuration of this type.

Having an airplane slowly hunt or change altitude during cruise is extremely annoying, and generally becomes worse in rough air. Such changes may be partially the result of a poor longitudinal trim system. Whether normally located on the fuselage or in the T-tail position atop the fin, I prefer a fixed or trim adjustable stabilizer to either the all-movable or V-tail designs. It has been my experience that the all-movable horizontal tail is subject to annoying trim changes with slight variations in airflow and also with slight wear and play in the control and trim system linkages, while the V tail causes both vertical and lateral displacement with each gust as previously noted. Quite likely the fixed stabilizer is a major reason for the solid feel and resulting popularity of the Cessna line of single-engine 172 Skyhawk aircraft that are so stable in flight.

The longitudinal trim system connecting the trim wheel with the elevator tab must have virtually no cable stretch or linkage play if the desired level of positive trim response is to be realized. We all appreciate a stable airplane that willingly trims to a 20 ft/min adjustment—and it is also a much easier and safer airplane to fly when the going gets a bit tight.

SUMMARY ITEMS

Stepping back a bit and putting together all these important parameters for cruising flight, we find that besides meeting FAA design and flight demonstration requirements the design group also had to put together an airplane acceptable to the sales department and one most satisfying to the new owners. To accomplish this for cruising flight requires a design that:

1. Has comparatively low drag to provide timesaving, efficient transportation for the powerplant used and the production cost.

2. Has ample and positive stability about all three principal flight axes—pitch, roll, and yaw. This requires the proper relationship and design compromise with the center of gravity, tail arm, horizontal and vertical tail areas, and the dihedral angle, as demonstrated in flight by (a) return to trim attitude when displaced, (b) positive control forces which steadily increase with control displacement from the trimmed position, and (c) maintaining an assigned altitude and heading in smooth air.

3. Has small trim displacement and elevator load changes with power.

4. Has an effective, responsive, and positive longitudinal trim control system.

5. Probably most important of all, has a powerplant that keeps operating (as will be covered at length in Chap. 11).

REFERENCES   7.1   Sidney M. Harmon, *Comparison of Fixed Stabilizer, Adjustable Stabilizer, and All-Movable Horizontal Tails.* NACA Report L-195, October 1946.

7.2   (a) David B. Thurston, "T Tails—Past, Present, and Future," *Aero Magazine,* May–June 1973.

(b) Donald R. Riley, *Effect of Horizontal Tail Span and Vertical Location on the Aerodynamic Characteristics of an Unswept Tail Assembly in Slideslip.* NACA TN 2907 (later issued as TR 1171), February 1953.

(c) R. R. Tumlinson and H. L. Walter, *T-Tail Aerodynamics of the Super King Air.* SAE Report 740370, April 1974.

7.3   FAR 23.427 (a) and (b).

7.4   FAR 23.427 (c).

7.5   (a) James S. Bowman, Jr., and Sanger M. Burk, Jr., *Stall/Spin Research Status Report.* NASA research contained in SAE Report 740354.

(b) James S. Bowman, Jr., *Summary of Spin Technology as related to Light General Aviation Airplanes.* NASA TN D-6575, December 1971.

8 APPROACH CONDITIONS

Although not listed with the primary operational criteria of page 55, the approach phase of flight presents some interesting design problems that are worth exploring.

Stability and controllability are all important at this time and should remain fairly constant over a broad range of relatively low airspeeds—assisted by some use of flaps as required. As a general rule, the final leg approach speed should be about 1.3 times configuration stalling speed in smooth air, and increased to 1.4 times stalling speed for rough air. But within these limits, the aircraft designer has considerable control over approach flight characteristics.

When entering final an airplane is flying near point G of Fig. 8-1, and may be operating at speeds below point J during power-assisted STOL approaches. Note that if the airplane is allowed to go below speed N there will be no excess power available for altitude or attitude recovery. This is shown at point K, where no gap exists between the power-required and power-available curves; this means no power reserve remains for climb recovery if the approach is too low or too short.

Flaps are of assistance here if they are the high lift slotted or Fowler types previously reviewed. These flaps have the effect of reducing approach speed for a given rate of descent, or conversely, increasing the rate of descent for a given approach speed. Further, they provide considerably more control flexibility with power to maintain the required glidepath.

Such desirable flight characteristics are clearly shown by Fig. 8-2, which is based upon actual flight data observed with a 150-hp Cherokee. With power and flap deflection stabilized for each curve, a direct reading was obtained for descent rate vs. airspeed, showing that for a fixed throttle setting and airplane configuration *airspeed was controlled by the stick.* Which is to say, *airspeed was controlled by the angle of attack* (pitch trim attitude) *of the airplane.*

These curves also demonstrate the relationship between *stick and throttle* in controlling the airplane, particularly at low speed. While I do not wish to join the FAA-industry–flight instruction merry-go-round about the effect of stick vs.

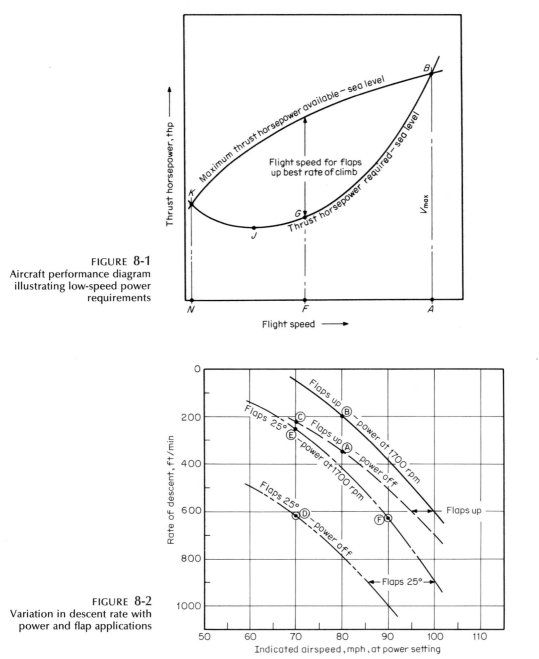

FIGURE **8-1**
Aircraft performance diagram
illustrating low-speed power
requirements

FIGURE **8-2**
Variation in descent rate with
power and flap applications

throttle on altitude and airspeed, it is quite obvious from Fig. 8-2 that both glidepath speed and slope corrections may be controlled through power changes. So the frequently expressed theorem that altitude is controlled by the stick should not be considered celestial law. In fact, I am sure most experienced pilots use both stick and throttle to apply critical glidepath corrections

—and probably without realizing which is applied at just what moment or in what sequence.

Note how the rate of descent may be varied along the approach path. In fact, descent rate *decreases* with airspeed right down to the stall—flaps up or down. This characteristic is common to all conventional aircraft and explains how a pilot tends to stretch a glide right into a full stall condition. Such accidents are most likely to occur in aircraft having small variations in control feel or trim attitude (angle of attack) with decreasing airspeed. There is an old theorem, "Make the pilot work hard to get into trouble," which means that stick forces should materially increase as the stall is approached. The modern Cardinal 177B does an excellent job of fulfilling this requirement, as will be well realized during the landing flare if elevator trim has been neglected. And this is the way the elevator control should be designed for safety.

By way of graphically depicting the control options available, Fig. 8-2 shows that if you are a bit low in the approach and coming in power "off," flaps up at 80 mph at point *A*, adding a bit of power to 1700 rpm will decrease the rate of descent from 350 ft/min to 200 ft/min per point *B*. This is *controlling altitude with the throttle*. However, there is another alternative just mentioned— pulling back on the stick. Slowing from 80 to 70 mph, power "off" and flaps retracted, would reduce the rate of descent from 350 ft/min at *A* to 220 ft/min at *C*. Unfortunately, attempting to further reduce descent rate by pulling back some more will eventually terminate the approach in a spin.

The 25° flap position curves, power "on" and power "off," show the same general relationship but because of flap drag have greater rates of descent for corresponding airspeed and power settings. For example, at point *D* the airplane is flying 70 mph at a power "off" descent rate of 620 ft/min with 25° flap extension. If things look marginally short, adding power to 1700 rpm will flatten the path to a descent rate of only 250 ft/min per point *E*. Some elevator adjustment will be necessary to maintain 70 mph, but this can occur relatively slowly in modern aircraft—and possibly due to gust conditions it may be considered desirable to approach a bit over 70 mph. Again, if the speed is reduced too much, a stall can occur with flaps extended, although the nose-high attitude and soft controls accompanying low-speed operation with flaps down should provide considerably more warning than when the flaps are up. This is another good reason for using flaps during final approach.

For the preceding reasons, the approach should be controlled by throttle for descent rate and altitude, and by stick and throttle for the speed desired. For example, if gusty conditions exist in the landing area, it would be advisable to increase speed to 1.4 times stalling speed while remaining at the desired descent rate. As shown by Fig. 8-2 the 70 mph, 620 ft/min power "off" descent rate of point *D* can be increased by a safe 20 mph margin to 90 mph at point *F* by adding power to 1700 rpm. Of course, some elevator trim may also be required.

Figure 8-2 shows what can and does happen if the stick alone is used to control altitude and glidepath slope (rate of descent) until aircraft attitude takes over and a stall/spin condition terminates the flight. Remembering these speed and power relationships should improve everyone's approach technique, especially in rough air.

While discussing the use of flaps, some mention should be made of the pitch trim changes resulting from flap deflection, particularly since this effect could become quite critical if not eliminated from the final design. By way of a somewhat simplified explanation, the effect of high lift flaps may be considered similar to increased lift acting along the wing trailing edge. As a result, flaps act to rotate the wing (and the entire airplane) nose-down about a spanwise pitch axis. In order to keep the airplane in a level flight attitude this nose-down moment must be balanced by increased tail load and elevator forces.

Because of flap-induced airflow changes over the horizontal tail some aircraft, the Piper Aztec and Beech Baron for example, tend to trim nose-up when flaps are lowered. So the type and amount of pitch change with flap deflection may vary from one airplane to another, making doubly important the need to design all aircraft for minimum trim changes with flap operation—both down and up.

If these unbalanced trim moments are large, the designer has three principal ways to reduce their effect upon flight handling characteristics: (1) a responsive elevator trim system capable of handling the unbalanced moments at maximum flap deflection; (2) interconnecting the flap actuation and elevator trim systems to provide more nose-up trim as the flaps go down, and (3) limiting the amount of flap deflection.

The first method is the simplest and most effective if it can be realized. At any rate, regardless of the design approach used to control flap deflection moments, flight tests should demonstrate that control forces can be handled when the flaps are fully down and the trimming system disconnected.

Getting back to Fig. 8-1, when approach speeds drop below the minimum power-required point J, the power necessary to maintain level flight rises rapidly with decreasing airspeed. This seems contrary to the basic laws of physics, since more power is apparently required to do less work. However, in the low-speed range between points J and K the airplane flight attitude (angle of attack) is such that total drag rises very rapidly and soon equals or exceeds the thrust available.

This region is known as the *back side of the power curve* or the *region of reversed command* because more and more power is required to maintain control (command) at lower and lower airspeeds. It is a good flight area to avoid and generally no problem for our relatively low-speed private aircraft. However, you can find yourself there when practicing STOL approaches, particularly in one of the cleaner twin-engine aircraft.

Fortunately, there is an instrument offering increased safety in this region: the angle of attack (AOA) indicator. By displaying acceptable ranges of angle of attack for optimum cruise as well as various flight conditions, the AOA indicator actually adds economy as well as safety to aircraft operation. Unfortunately, for some reason this flight instrument seems neglected by private aviation; I'm sure approach accidents would decrease in direct proportion to AOA indicator acceptance and believe this device should be given more attention. For readers interested in pursuing this flight regime a bit further, Ref. 8.1 includes material covering the subjects of stick vs. throttle, STOL operation, and flying angle of attack.

Similar to the requirements for climbing and cruising flight, a positive level of static and dynamic stability about all three axes is important during the approach, particularly in IFR conditions. In addition, moderate control forces should be well balanced between all surfaces, adverse yaw should be very low, control response should be high right into the stall, and the longitudinal trim system should be effective.

While the design procedures required to obtain all these desirable qualities have been reviewed in the preceding chapters and will not be detailed again, it is interesting to note that FAA flight test approval requires approach speed demonstrations proving these characteristics are present at acceptable levels throughout the desired gross weight center-of-gravity range. To mention just two of many such tests: (1) a balked landing (missed approach) full-power climb-out without touching elevator trim is required to show whether a new design has undesirably high or reversing stick forces; (2) ever tightening circular flight in approach trim, plus sudden application of full power, represents a negligent approach or an avoidance maneuver which must be accomplished without having the airplane fall out of the sky or place extreme demands upon the pilot. Of course, such tests place considerable demand upon the airplane designer, and frequently they result in modifications to the control and trim systems along with control surface area, planform, and contour changes.

Responsive engine controls are essential during the descent and landing phases, probably to a greater degree than for any other nonacrobatic phase of flight. When power is needed, it is needed *now*, and not one-half second later due to a sloppy linkage or cable control. For similar reasons, effective carburetor heat control and response is also mandatory. In view of the number of accidents attributed to engine icing during the approach, possibly some consideration should be given to having carburetor heat linked to fully retarded throttle operation, as well as being independently controlled. This might eliminate another source of "pilot error" by automatically applying carb heat whenever the throttle is pulled all the way back.

In summary, all desirable approach characteristics are brought into sharp focus during IFR flight. Positive stability, minimum adverse yaw, effective surface controls with rapid response to very small deflections, a glidepath control system such as wing flaps, and positive engine controls all become very critical in the fog—and serve to immediately separate good, bad, and indifferent IFR designs. In addition, the ability of higher-performance aircraft to establish a steady, standard descent rate approximating 500 ft/min by lowering flaps, gear, or both when passing over the final approach fix is a source of great comfort and workload relief when conditions are piling up in the cockpit. The Piper Cherokee Arrow does this quite well with gear extension and without throttle adjustment at an approach speed of about 100 knots.

*   *   *

Since landing control and handling criteria were quite thoroughly reviewed during the crosswind discussion of Chap. 5, they will not be further considered here; so our flight profile now closes with this brief approach review. In addition to providing some design background for the various segments of routine flight, it is hoped these chapters have served to improve every reader's flight technique to some degree.

We next proceed to the important design and flight safety area of weight and balance control, followed by a performance summary section. Subsequent chapters include general and detail design information items that should be of interest and benefit to every pilot and aircraft owner.

REFERENCES 8.1 (a) Dan Manningham, "Flying Angle of Attack," *Business and Commercial Aviation Magazine,* September 1975, p. 54.

(b) FAA, *Airplane Flight Characteristics and Techniques Associated with Operation in the Region of Reversed Command.* Advisory Circular AC No. 61-50A, March 15, 1976.

9 USEFUL LOAD AND WEIGHT CONTROL

Useful load, the last of our basic operational criteria, can make or break a new airplane intended for the serious business and pleasure markets. In view of this, every new design must provide as much flight performance and safety as possible at minimum empty weight.

To clarify weight and balance terminology before briefly covering the related calculations, it is important that the following weight conditions be clearly understood: *useful load* is the difference between the *empty weight* established when the airplane was completed by the manufacturer and the *gross weight,* which is the maximum legal flight weight.

Since empty weight consists of the basic airframe, engine, instruments, and residual (undrainable) fuel and oil required to satisfy the minimum FAA airworthiness requirements, everything added until gross weight is reached is considered useful load. This means any primary blind or full IFR instrumentation, radios, antennas, wheel fairings, special interior trim, or special equipment (other than the mandatory emergency locator transmitter) will actually be part of the fixed useful load. Such permanent additions rapidly reduce the disposable weight that can be carried in the form of oil, fuel, passengers, and baggage. Just a few options or modifications can have a critical effect upon duration and range, IFR suitability, and passenger-carrying ability—all basic parameters directly related to aircraft utility and acceptance.

In view of this, some helpful background will be gained from reviewing design procedures followed in developing the maximum useful load for a new airplane.

EMPTY WEIGHT AND PRELIMINARY DESIGN

The basic secret is getting and keeping a low empty weight for the powerplant size and design load factors required. And low empty weight means low airframe weight, which can only be realized through efficient structural analysis and system design.

For example, if the wing skin could apparently be 0.032 in. thick according to stress analysis, the designer should not use 0.040-in. skin. In fact, if time and money permit, static test of a wing panel may prove that skin 0.025 in. thick is adequate. However, some trade-off may be required for this decision. If the airplane is intended for bush-country operation, it might be more prudent to leave the skin at 0.032 in. or go to 0.040 in. for a rugged structure, saving weight in some other area such as canvas sling seats instead of more comfortable uphol-stered ones—this is utility appeal. On the other hand, if the airplane is intended for the more gently operated private market, thin skin could be used and the weight saved converted to useful load or deluxe upholstered seating, both of which carry sales appeal.

Unfortunately, low structural weight usually requires thin skins and many small stiffeners and frames. This type of airframe design is obviously contrary to efficient, low-cost production and would result in an excessively expensive air-plane. This approach may be acceptable for special experimental work to prove a new design concept, but certainly is not suitable for mass production as practiced by the private aircraft industry. So again the designer must compro-mise, this time balancing the weight empty goal with production labor-hour re-quirements. This takes time and partially explains why aircraft are expensive. Total engineering costs must be prorated over a predetermined number of pro-duction units, each carrying a portion of the design and development costs.

Somewhere around the first third of the total design period, someone always comes forward with a requirement for more horsepower; the theory apparently being that the final design can be made available for the same price if a larger engine is incorporated early enough in the design program. This is never so, since a larger engine and propeller cost and weigh more just sitting in their shipping containers; and they require heavier structure to support their own greater weight as well as the increased flight speed they are supposed to deliver.

It is true, however, that if a new design is to be eventually developed into a line of similar aircraft having increased power and useful load, overall weight can be saved if the initial structure is designed with growth in mind. Although some penalty will be incurred by the lower-powered, basic configuration, sub-sequent designs of higher power will be increasingly efficient. For a given gross weight and operational requirement, it is always less costly in both structural weight and final price to build strength into a basic structure rather than adding reinforcements and reworking the airframe at a later date.

Since close weight control is essential during the actual detail design phase, accurate weight estimates for the various airframe components must be devel-oped during the preliminary design study program. These weights establish an empty weight target for each new airplane and should be carefully monitored right through certification.

Wing, fuselage or hull, tail surface, equipment, powerplant, and landing gear weights will vary as a percentage of the total design weight, generally in accordance with operational use. As examples, the greater the applied load

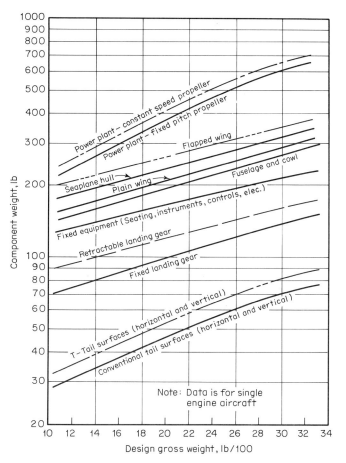

FIGURE **9-1**
FAR Part 23 aircraft component weight vs. design gross
weight

factor requirements, the heavier the structure; the more equipment necessary to perform a service such as agricultural spraying, the larger the percentage of empty weight required for equipment items; and so forth. Figure 9-1, prepared from detail weights for many single-engine aircraft designed to recent FAR Part 23 requirements, is included to serve as a preliminary design guide for basic weight estimation and to indicate how structural weight increases with gross weight.

Once these various weights have been more or less established, it is necessary to properly arrange the different airframe and useful load items so the airplane will balance as desired throughout all flight and loading conditions. These preliminary weight and balance calculations are usually presented in a form similar to Fig. 9-2. This is known as the preliminary weight and balance summary and is prepared the same as the individual calculations and log entry necessary any time equipment items are added to or removed from a licensed airplane.

Every designer is faced with the problem of providing center-of-gravity

WEIGHT AND BALANCE SUMMARY

Page _____

L.E. MAC @ Sta. _____

MAC = _____ in.

Configuration _____

No.	Item	Wt, lbs	Horizontal arm, In.	Horizontal moment in. lbs	Vertical arm, in.	Vertical moment, in.-lbs	Alternate wt, lbs	Arm, in.	Moment, in.-lbs
1	Fuselage/hull								
2	Windshield								
3	Sliding canopy or doors								
4	Aft canopy or doors								
5	Horizontal tail								
6	Vertical tail								
7	Wing								
8	Engine								
9	Engine mount								
10	Cowl and exhaust system								
12	Propeller, gov. and spinner								
13	Carburetor air box								
14	Fuel & oil system								
15	Surface controls								
16	Furnishings								
17	Instruments								
18	Electrical system								

FIGURE **9-2**
Weight and balance summary form

19	Main gear (retractable)									
20	Nosewheel (retractable)									
21	Hydraulic system									
22	Battery inst.									
23	Reserve fuel and oil									
	Empty weight									
24	Pilot and passengers									
25	Rear passengers									
26	Oil (____ qt)									
27	Fuel (____ gal)									
28	Accessories (radio, etc.)									
29	Baggage									
	Useful load									
	Gross weight									

$$\left(\frac{\text{Horizontal arm—L.E. MAC Sta.}}{\text{MAC}}\right) = \begin{array}{l}\text{c.g. location from MAC leading edge in fraction}\\ \text{of MAC (multiplying by 100 gives percent MAC)}\end{array}$$

Report no. _____

Date _____

Model _____

Cont _____

limits that will accept all or most loading conditions necessary for convenient and useful operation. Since flight tests must demonstrate the existence of adequate stability at the extremes of the center-of-gravity envelope, the desired center-of-gravity range may influence horizontal and vertical tail surface design. Because longitudinal, lateral, and directional stability are all affected by tail surface area, the need for a center-of-gravity range extending far aft may result in increased tail arm or areas, or both, as reviewed in Chap. 7.

A desirable center-of-gravity range for powered aircraft would extend from about 15 to 28 percent of the average wing chord, more correctly referred to as the *mean aerodynamic chord* or *MAC*. Sailplane designers prefer the center of gravity further aft to keep the horizontal tail relatively unloaded and so reduce tail area and drag, but such a center-of-gravity location is usually accompanied by neutral or slightly positive longitudinal stability and very low stick forces which are not acceptable for private aircraft operation.

As shown by Fig. 9-3, to obtain the fore and aft location of the center of gravity the weight of each component is multiplied by its distance (or arm) measured from a vertical reference line called the horizontal datum, to the center of gravity of the component. These horizontal moments are summarized on Fig. 9-2 where indicated. To obtain the horizontal center-of-gravity position the summary total of the horizontal moments is divided by the total weight of all the items used to compute the total horizontal moment, in this manner

$$\frac{\text{Summation of horizontal moments}}{\text{Total weight of items}} =$$

c.g. location from the vertical reference (or datum) line, in.

The vertical center-of-gravity position is determined by the same procedure except the arm to each weight item is measured from a horizontal base line, or vertical datum, and then summarized on the form of Fig. 9-2 alongside the horizontal moment values. The reference datum lines for these fairly simple but laborious calculations may be conveniently located as shown on Fig. 9-3; they will be clearly marked on the airplane side view included with every aircraft weight and balance statement.

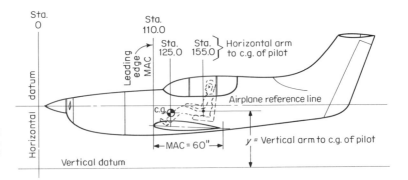

FIGURE 9-3
Horizontal and vertical datum line locations, with an example shown at the center of gravity for the pilot.

The center-of-gravity location as a percentage of the MAC is simply determined as follows. Again referring to Fig. 9-3, if the MAC leading edge is at station 110 and the MAC is 60 in., when the center of gravity is at station 125 we have

c.g. location referred to the MAC

$$= \frac{\text{(c.g. location)} - \text{(leading edge MAC station)}}{\text{MAC}}$$

$$= \frac{125 - 110}{60} = \frac{15}{60} = 0.25 \text{ (or 25\% MAC when multiplied by 100)}$$

The form for this calculation is shown at the bottom of Fig. 9-2.

After the balance range has been computed within acceptable limits, the center-of-gravity position should be closely monitored during the entire design period. A high level of weight supervision is mandatory to avoid added ballast or radical last minute changes in equipment location to preserve center-of-gravity limits. Unfortunately, the degree of weight control available is all too frequently insufficient to provide the direction required; and so some fast changing usually occurs immediately after the first weighing of a new design.

Active weight control frequently results in redesign of a structural component that is becoming too heavy. Although removal of excess weight may be costly in time, engineering, and materials, it is essential if the desired useful load is to be realized.

Current FAA requirements add to the empty weight all fuel and oil that cannot be readily drained on the ground or used in flight, whichever is greater. And the flight condition is the problem with modern aircraft having wing fuel tanks, because the amount of this residual fuel is determined during a steady slip or skid—again whichever maneuver results in more fuel remaining in the tanks when the fuel pressure begins to drop off. Since wing tanks are flat by nature, a considerable volume of fuel can be piled up away from the tank outlet. Most of this fuel could be used in normal, level flight, but the FAA considers as critical the possibility that someone will run dry during a steep slip or skid on final. This could happen if fuel is low or the tank on the slip side is in use, both of which conditions indicate poor fuel management but occur regularly.

Unfortunately, all residual fuel weight established by these rather extreme tests must be included in the certified weight statement as part of the empty weight—and, of course, the useful load is reduced accordingly. On some small aircraft this value approaches 30 lb, representing the weight of a good dual nav/com radio installation. As a result, increased attention must be paid to the detail design of fuel systems as will be further covered in Chap. 11.

USEFUL LOAD The weight and balance statement supplied with each airplane is obtained by actually weighing the completed airplane prior to delivery. These statements have legal value since they represent the licensed weight empty and are part of the FAA Certificate of Airworthiness requirements applying to every production airplane sold. This weight statement forms part of the airplane flight manual carried in new aircraft or is part of the owner's manual supplied with older airplanes. As such, the weight statement must remain in the airplane at all times.

As previously stated, the difference between the actual certified weight and the design gross weight represents the total allowable useful load. However, a good percentage of this value may have been permanently removed through installation of IFR instruments and avionics, special equipment additions, and structural repairs or modifications. What is left is the disposable load capacity available to carry oil, fuel, passengers, and baggage.

To make sure allowable gross weight or approved center-of-gravity limits are not exceeded, every pilot must be fully aware of the useful load being carried on each flight as well as the way this load is distributed in the airplane. Just because an airplane is advertised as having a 1000-lb useful load, do not assume 1000 lb can be added to that ship parked on the ramp. The Certified Weight and Balance Statement as amended to include current changes in equipment and airframe modifications is the sole source and authority for useful load capability.

At the risk of being accused of beating a dead horse, the danger of overloading must be mentioned again since private aircraft performance deteriorates rapidly with overload even under standard air conditions; to say nothing of performance losses under high ambient temperatures or pressure altitudes as covered in Chap. 6. A set of typical flight manual load and moment diagrams are shown in Figs. 9-4a and 9-4b.

FIGURE 9-4a
Chart of moment values for different useful load items. Procedure: (1) Find the point on the line corresponding to the flight loading and record the weight and moment in the table. (2) Locate the final weight and moment totals (example for 2112 pounds at 219,311 in.-lb) on the weight-moment chart of Figure 9-4b. The final weight and moment values must fall within the center of gravity limits established by the lines indicating forward or aft center of gravity.

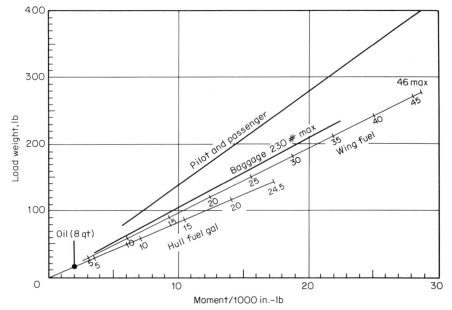

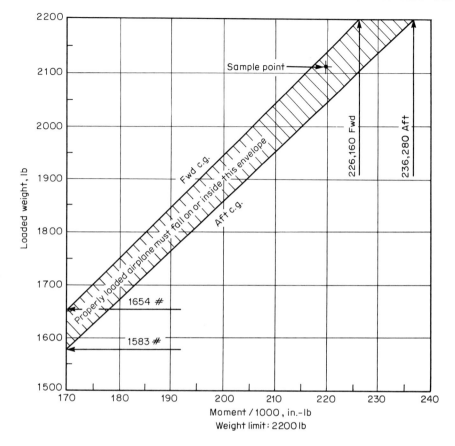

	Actual problem			Sample problem		
Load item	*Weight*	*Arm*	*Moment*	*Weight*	*Arm*	*Moment*
Licensed empty weight/equipment		—		1454	—	162,867
Oil (8 qt = 15 lb)		126		15	126	1,890
Wing fuel (46 gal = 276 lb)		103		180	103	18,540
Hull fuel (25.5 gal = 147 lb)		118		3	118	354
Pilot and passenger		71		340	71	24,140
Baggage		96		120	96	11,520
Totals				2112	103.8	219,311

FIGURE **9-4***b*

Chart locating total load and moment values vs. center of gravity limits. Flight must be conducted within the center of gravity limits shown at any weight.

Since stalling speed will vary directly with the square root of the weight ratio

$$V_2 = V_1 \sqrt{\frac{W_2}{W_1}}$$

for each 10 percent increase in gross weight the stalling speed increases by 5 percent. This means more than a 5 percent longer takeoff run assuming comparable ambient temperature and pressure conditions, followed by a decreased ROC.

As indicated by the climb formula of Chap. 6, climb rate is inversely proportional to aircraft weight, assuming no variation in climb drag with aircraft weight. This is shown by the ratio

$$ROC_2 = ROC_1 \left(\frac{W_1}{W_2}\right)$$

which roughly indicates a 10 percent reduction in climb for a 10 percent increase in weight. However, the airplane drag coefficient will increase slightly with weight since the wing assumes a greater angle of attack to carry the heavier weight; this in turn increases the induced drag as well as the overall drag resulting from increased fuselage angle of attack and interference drag. So our ROC actually decreases faster than the simple weight ratio indicates, and frequently catastrophically so.

Recent accident data published by the National Transportation Safety Board (Ref. 9.1) list 40 to 70 fatalities per year over the past few years directly attributable to poor weight and balance management. In a number of crashes Cherokee 140s or Cessna 150s were being flown up to 240 lb over gross, and in at least one instance the center of gravity was 2 in. aft of the approved rearward limit. Please take care—there is only so much a designer can do to make flying safe and enjoyable; and overloading has never been an approved means of increasing useful load.

WEIGHT AND BALANCE CONTROL RECORDS

The aircraft weight and balance statement supplied by the manufacturer must be modified each time new items of fixed equipment or structure are added to the airplane, as well as whenever any previously installed equipment or structure is permanently removed.

This is a legal requirement of FAA flight regulations intended to ensure that persons using an airplane will have the current weight and balance values at their fingertips along with the aircraft flight manual. As a result, the owner of every airplane is held responsible for maintaining current weight and balance control records; and this applies whether the owner is an individual, a fixed-base operator, or any type of corporation.

Actually, maintaining these records is quite a simple procedure and one that is frequently practiced by every FAA licensed aircraft mechanic. However, you may be able to save some time and money by working out your own weight statements whenever any changes are necessary. If nothing else, the following brief example will provide the background necessary to check the mechanic's calculations the next time some modifications are required.

The basic calculations for adding a second set of radios plus another communications antenna may be tabulated in this manner:

	Weight	Horizontal arm	Moment
1. Aircraft weight empty plus fixed equipment (per factory weight & balance)	1435#	86.06''	123,500''#
2. Add second nav/com	15	65.50	983
3. Second com antenna	3	120.75	362
4. Flight line weight & c.g.	1453#	85.92''	124,845''#

The values across line (1) are taken right from the current weight and balance statement. These numbers will be those supplied by the factory for an airplane not previously modified, or the latest weight log totals for a ship that has some changes. At any rate, to these basic totals we add the weight and horizontal location of the new nav/com as shown in line (2). This location will be the horizontal distance, or horizontal arm, from the aircraft horizontal datum to the center of gravity of the installed avionics equipment per Fig. 9-5.

The equipment manufacturer will show this center-of-gravity position in the installation data, and presumably you or your mechanic can measure from some known airframe reference point to the equipment center of gravity. Usually the aircraft flight manual or owner's manual will list some structural reference station such as a door frame or wing leading edge for this very purpose.

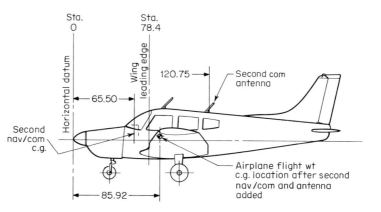

FIGURE 9-5
Location of items for the weight and balance calculations required by the addition of a second nav/com radio and antenna.

Multiplying the nav/com weight by this arm provides the moment for the installed radio, 983 in.-lb for our example. In similar manner the second antenna is added into the weight and balance calculation as shown by line (3). Note that fractions of in.-lb have not been included when they are less than 0.50; if 0.50 or larger the next full in.-lb is used. For our installation 982.5 becomes 983 in.-lb and 362.25 is listed as 362 in.-lb.

We next add the weights on line (4) and find the new total of 1453 lb. This is properly referred to as the *flight line weight* since this value represents the *equipped* empty weight of the airplane before the *disposable* useful load has been added. Note, however, that there will now be 18 lb less capacity available for passengers, baggage, fuel, and oil—representing the decrease in disposable useful load due to the added fixed weight of the avionics equipment.

This equipped empty weight is frequently referred to as the airplane empty weight, which is acceptable practice as long as no one believes it is the same as the basic airplane empty weight before fixed equipment has been added. This would mean the full useful load could be carried, and such a misunderstanding could be the first step toward an overload accident. As a matter of good practice, inspect the airplane weight and balance log whenever you first fly any airplane; and make sure you have the latest weight statement. Frequently the weight changes go on for many pages with no reference to the current final statement—a bad practice, but a very common one.

As a final step, adding up the new moment total and dividing by the new weight of 1453 lb gives us the new horizontal arm of 85.92 in. from the datum to the flight line weight center-of-gravity position. Note that the airplane center of gravity has moved forward a bit from the original 86.06-in. location, as would be expected when equipment is added ahead of the instrument panel.

Regardless of the type of weight change, the above procedure will supply the required answer. If any reader wishes to study this subject at greater length, Ref. 9.1 and 9.2 carry more complex weight and balance examples of varied detail. Reference 9.2 is also of particular interest if structural modifications or repairs must be documented in accordance with FAA requirements.

In conclusion, I believe one of the most important facts an aircraft owner can remember is that from the time a well-designed airplane first flies, everything done to it will contribute to increased empty weight, reduced performance, and greater cost. This also holds true throughout the life of the airplane; like cruising boats and human beings, aircraft seem to grow heavier with age. And since useful load as well as takeoff and climb performance decrease with increased weight, both utility and safety suffer from every addition of unnecessary fixed weight.

REFERENCES   **9.1**   *Weight and Balance: An Important Safety Consideration for the General Aviation Pilot.* National Transportation Safety Board report NTSB-PAM-74-1, December 18, 1974.

9.2 Pages 243–265 of FAA Advisory Circular AC 43.13-1A, *Acceptable Methods, Techniques, and Practices—Aircraft Inspection and Repair.* Available from:

Superintendent of Documents
U.S. Government Printing Office
Washington, D. C.   20402

Order Publication FAA AC No. 43.13-1A; stock No. 5011-0058.   Price $3.10 postpaid.

10 OPERATING PERFORMANCE

In one sense, this chapter is a summary of all preceding design data; by weighting the various parameters for each phase of flight it is possible to show their individual importance for aircraft design. Since this information will be of interest to pilots as well as those readers particularly concerned with aircraft design, it will be presented before beginning a more detailed performance review.

THE ORDER OF MERIT

The effect of eleven parameters—ground handling, power, constant speed propeller, wing area, aspect ratio, flaps, low-speed control, stability, drag, responsive engine controls, and weight—have been valued for each phase of flight. The more important the contribution to safe and acceptable flight characteristics for the specific operation, the higher the unit value, with 10 being the maximum assigned for any item.

Admittedly the priority listings may appear arbitrary to some readers, in which case the values can be reassigned and a new set of priorities developed; however, they will probably end up in quite similar final order. From a design standpoint the relative values used in the tabulation of Fig. 10-1 apply to current production aircraft and so represent a working demonstration of modern private aircraft design practice.

Although all of these parameters have some obvious value and importance to aircraft design, by adding together the priority values for each one in each phase of flight we obtain a listing of relative design merit. When arranged in the order of decreasing totals, and therefore importance to modern aircraft design, we find the following results:

1. Stable airplane—37 points
2. Horsepower—36
3. Constant speed propeller—35

4. Low weight—35

5. Low-speed control—29

6. Large wing area (comparatively low-wing loading)—27

7. Flaps—25

8. Clean airplane (low-drag design)—24

9. Ease of ground handling—20

10. High aspect ratio—18

11. Responsive engine controls—14

Small wing area necessary for very high-speed cruise represents a special condition not relevant to normal private aircraft operation, and so finished well down in the rating with only 6 points. However, if speed were paramount, small wing area would be weighted differently and probably replace large wing area at the number 4 position. The point being that data of Fig. 10-1 were pre-

FIGURE **10-1**
Design parameters for flight conditions, FAR Part 23 airplane, weighted by importance to each phase of flight operation.

Takeoff	Climb	Cruise	Service ceiling	Descent and approach	Landing
10. Ease of ground handling and control (tricycle gear)	10. Thrust hp margin (constant speed propeller for maximum thrust)	10. Stable airplane	10. Maximum hp margin (turbo-charged engine for maximum alt. perf.)	10. Stable airplane with minimum adverse yaw	10. Tricycle landing gear
9. Thrust (hp and constant speed propeller for maximum thrust)	9. Low weight	9. Clean airplane	9. Constant speed propeller to deliver maximum thrust	9. Flaps for speed and descent control (or use of spoilers, lowering landing gear, etc. to control drag value)	9. Aileron, rudder and elevator control systems that are responsive at low speed with minimum adverse yaw
8. Low speed rudder and aileron control (elevator also for all seaplanes and STOL landplanes)	8. Clean airplane to reduce climb drag (faired landing gear, clean cowl, wing root fillets, etc.)	8. Constant speed propeller to provide takeoff and climb without penalizing cruise performance	8. Clean airplane to reduce drag	8. Responsive surface and engine control systems	8. Combination of wing area and flaps adequate to provide low landing speed (particularly important for seaplanes and short-field operations)
7. Low weight	7. Large wing area and high aspect ratio to reduce C_L and induced drag:	7. Relatively high power to permit lower rpm and quieter cruise	7. Low weight	7. For IFR work, the ability to rapidly set up a 500 fpm rate of descent over the approach fix	7. Low weight for low landing speed
6. Wing area and flaps to provide low takeoff speed (particularly important for seaplanes and short-field operation)	$$C_{D_i} = \frac{C_L^2}{\pi AR}$$ C_L varies inversely with S_w and therefore C_{D_i} decreases as the square of lift reduction— also C_{D_i} decreases with AR increase	6. Small wing area and high aspect ratio to reduce parasite and induced drags (a real trade-off required here)	6. Stable airplane (cannot attain maximum ceiling if control surfaces must be constantly operated to maintain desired flight path)		6. Responsive engine and throttle system for control in gusty air or aborted landing
	6. Stable airplane	5. Low weight	5. Large wing area and high AR to reduce C_L and induced drag		5. Stable airplane
			4. Responsive surface controls with minimum adverse yaw		

pared for a typical FAR Part 23 private aircraft design, and should be considered primarily for that class of airplane.

As would be expected under current FAA design requirements and in consideration of flight safety, a stable airplane configuration is of primary importance; this parameter designs the tail, wing, and control surfaces.

While previous chapters may have pointed out the cost and weight penalties associated with horsepower, as will be recalled these criticisms referred to the addition of power to overcome design deficiencies such as heavy empty weight or insufficient wing area for acceptable takeoff and climb performance. The correct selection and use of horsepower, or more properly thrust horsepower, is important to every basic airplane design since it applies in some manner to all phases of powered-aircraft flight. In view of this, it is not surprising to find powerplant parameters tied for second- and third-place importance in our evaluation.

Low weight appears high on the list since it is not only an indication of design and structural efficiency but also is important throughout the entire life of the airplane. For a given task, the lighter airplane should cost less and will be more economical to operate.

Comparatively large wing area, low-speed control, and flaps all apply to safe low-speed operation from small airports and under crosswind conditions. So we find these criteria grouped to form a middle of merit rating that binds together the primary and secondary criteria, items 1 through 4 and 8 through 11 respectively.

It is surprising, though, to find low drag in eighth position, 13 points out of first and 10 points above last place. By way of explanation, although low drag is desirable throughout the entire flight spectrum, it is only critical for three of the six flight conditions listed. Because of the minimum effect of drag at low speeds, this parameter does not seriously enter takeoff or landing considerations for private aircraft design, while a high drag value may actually be desirable during the approach. And of the three conditions affected by low drag, only climb and cruise are important; service ceiling is rarely called for in our type of flying, although it would deserve more consideration for mountainous areas of the West, Canada, South America, and Africa.

High aspect ratio reduces induced drag during climb and cruise, releasing additional power to increase climb rate and cruising speed. However, in return for this gain we have a heavier wing structure which increases the empty weight, and also the gross weight if a constant useful load value is to be maintained. And since climb varies inversely with weight, we lose to increased weight some of the climb provided by increased aspect ratio.

Ease of ground handling and responsive engine controls are self-evident, although I'm sure that many student pilots and their instructors frequently consider ground handling of primary importance. Again as a reminder, the merit order of these design parameters is based upon requirements for a private airplane design and so is not weighted toward trainer use. But you can easily rearrange the priorities for such an application yourself to determine how the resulting airplane should be designed. Stability would still be paramount, but power would drop in importance while ground handling and low-speed control advance in rating. The result: a Cessna 150, Cherokee 140, or perhaps the old Piper Cub and Aeronca Champ.

WHAT PRICE
PERFORMANCE?

Competitive aircraft performance is just as important as efficient structural design and handling characteristics in determining acceptance. If an airplane does not climb or cruise at least on a level with others of its class, it will not be successful regardless of how structurally sound it may be or how easily maneuvered. Since the performance diagram ranks in equal importance with the V- n envelope (Fig. 2-2) in detailing aircraft characteristics, we will briefly review the information carried by the performance curves of Fig. 10-2.

At speed A the power-available and power-required curves are seen to intersect at B. This is the maximum speed possible at the altitude for which the curves have been derived. There is no more power available, so the airplane can not exceed power-required speed A. When the engine output is reduced to 65 percent power, the intersection at D indicates cruising speed C is obtained at this power setting. The thrust horsepower available between D and E represents the power available to climb at speed C when the airplane is trimmed for 65 percent power cruising flight at point D.

These power-available and power-required curves are usually prepared for sea-level standard conditions. However, for performance comparison, power-required values may be given for other altitudes similar to the 7000-ft curve through point D', while additional power-available curves would be prepared for supercharged engine operation. Although considerable calculation is required to obtain power required vs. speed, this curve is sometimes presented for more than one weight or design configuration if such data are necessary for important operating conditions (such as switching from faired landing gear to seaplane floats, which affect both weight and drag).

Although detail performance calculations are quite lengthy and so will not be covered in this book, Ref. 10.4 contains a list of suitably comprehensive sources for those interested in developing performance diagrams.

The various design features affecting the climb, cruise, and approach phases of the performance diagram have been thoroughly covered in Chap. 6, 7, and 8 and will not be mentioned again. As a summary reference, Fig. 10-1 shows the effect of various options upon aircraft performance. High speed requires high power, low wing area, and low drag; climb is improved by adding high aspect ratio wing area, increasing power, reducing weight, or by all three; and cruising speed is related to moderate wing area, an efficient powerplant, and low drag. Since drag is mentioned with every reference to efficient operation and has not previously been covered in depth, this would seem a good point to

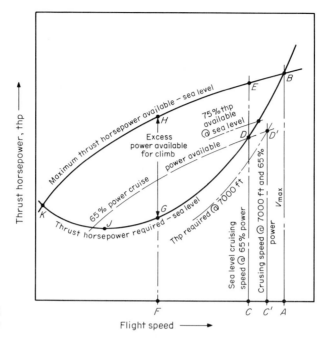

FIGURE 10-2
Aircraft performance diagram

review drag causes and cures, particularly since drag is basic to all performance evaluation, comparison, and improvement.

DRAG Like death and taxes, drag is always with us—and just about as welcome. Regardless of flight attitude or speed, drag forces are continually exerting some effect upon aircraft performance and handling characteristics. In fact it is virtually impossible to discuss drag without also considering performance and handling qualities.

As we realize only too well today, automobiles consume more fuel per mile at higher speeds because of increased bearing, drive system, and engine friction (internal drag) plus aerodynamic resistance (external drag), both of which must be overcome as speed increases. In similar manner, displacement boats require tremendous amounts of power to increase speed slightly beyond a design speed established by waterline length—and depending upon hull shape, may not be able to exceed this speed regardless of the power available.

Since aircraft have three degrees of freedom of motion, drag affects their operation in more ways than it does surface vehicles. In addition to lowering cruising and top speeds and thereby increasing fuel consumption, drag, as previously discussed, also decreases ROC and increases the rate of descent during approach; this in turn increases the skill required during landing flareout. Although an increased rate of descent may be desirable on final approach when accompanied by low speed, the other negative characteristics certainly arc not welcome and occasionally become the deciding factor between a successful flight and a statistic.

Similar to airplane weight, drag also has an insidious way of increasing with age; everything added onto or into the airplane contributes just a bit more additional drag. Because drag is so closely related to aircraft operation, we will consider in detail the basic types, sources, and causes of airplane drag. In addition, some performance improvement features will be noted; these should benefit your airplane and would be incorporated into any new design.

WHAT CAUSES DRAG? Airplane drag is generally considered the sum of two different types: parasite drag and induced drag. From the designer's standpoint it is preferable to separate drag into three components: parasite drag, wing profile drag (from airfoil section drag), and induced drag. Isolating wing profile drag from parasite drag gives the designer a direct indication of the effect of various airfoil sections upon cruising flight. This is an important consideration since the lower the wing profile drag at cruise lift coefficients, the greater the speed range from stall to maximum level flight speed and the more efficient the airplane.

Now just what causes these types of drag? Their source falls into eight general categories as shown below. The total value of drag types 1 through 7 represents the airplane *parasite drag* and this sum plus the *induced drag* will be the *airplane total drag*. The various total drag components are

1. *Flat plate drag.* Due to the frontal area of the airplane. The cross-sectional area of the fuselage (or hull) and other structural members such as exposed landing gear, floats, or skis may be considered as so much barn door area parasite drag and explains why high-performance aircraft are designed to have minimum frontal area and retractable landing gear.

2. *Skin friction drag.* Air passing over a surface creates drag in proportion to the amount of exposed area and the roughness of the surface. For this reason high-performance aircraft are designed with flush-riveted fuselages, tail surfaces, and wings (which are also of minimum surface area), while high-performance sailplanes are designed with minimum total surface area which is then hand-rubbed to a polished finish.

3. *Interference drag.* Caused by the breakdown of airflow over external structural and accessory items, such as landing gear struts, wheels and skis, wing and tail surface brace struts, flying wires, exposed bolt heads and nuts, drain fittings, radio antennas, and similar appendages. Frequently, this turbulent air flows over other parts of the airplane causing additional drag plus a possible reduction in surface effectiveness, control response, and aircraft stability.

4. *Separation drag.* Occurs from the overexpansion of air flowing over rough or poorly designed intersections and structural contours, such as abrupt changes in fuselage shape, the sharp break of hull lines at the transverse step

in a hull bottom, irregular contour of the wing leading edge and upper surface, the need for wing root area fillets, cowling flow breakdown, and protruding canopy shapes—to name a few causes.

Separation drag may frequently be of an alternating attaching/detaching flow type. This problem is particularly common with amphibians and can manifest itself as a most annoying cause of left/right hunting in yawed flight, with the airplane first heading off about 10° in one direction and then swinging back to a similar displacement in the opposite direction. The source of such a flow breakdown may be difficult to isolate and equally challenging to cure, but it is usually associated with flow from a cowling, wing root, landing gear, or strut intersection that changes the pressure field flowing over the vertical tail which in turn causes the oscillating flight.

5. *Pressure field drag.* Somewhat similar to a combination of interference and separation drags. When leakage flow occurs around or past poorly fitting doors, removable panels, rudders, elevators, ailerons, and flaps, the change in pressure due to flow separation or induced turbulence creates additional drag. This condition is probably most apparent as whistling doors (due to poor sealing) and leaking cowling panels; but flow leakage around control surfaces is probably a greater source of drag at higher speeds, and reduces control response at all speeds.

6. *Powerplant drag.* Represents the energy used to cool the cylinders and oil system, provide combustion air, and drive accessory items such as fuel, hydraulic, and vacuum pumps, prop governor, and alternator. The amount of cooling air required during climb will usually design the air inlet and exit areas. For higher-powered aircraft cooling flaps are frequently used to provide adequate outlet area during climb, and are then closed for minimum cooling drag during cruise flight.

7. *Wing-section profile drag.* Varies with the airfoil section selected by the designer. For maximum cruising range and for high-performance aircraft a low value of wing-section profile drag is desirable. The airfoil should be selected with low profile drag, low section moment (page 75), and gentle stall characteristics as primary considerations. As noted, by keeping the airfoil-section profile drag separate from the airplane total drag, the designer may readily determine the effect of different airfoils upon the operational characteristics of a new design. (See Ref. 6.3 for airfoil data.)

8. *Induced drag.* This category of airplane drag is particularly unique because it is composed of two complementary parameters that can also make or break a design from the moment the configuration is established. These are wing area and aspect ratio, and both affect the magnitude of induced drag. Since induced drag may represent one-half to two-thirds the airplane total drag value at climb speeds, the effect of wing area and aspect ratio will again be briefly noted.

At a given weight and speed, the lift coefficient necessary to sustain level flight will vary with wing area: the more wing area, the lower the required value of lift coefficient C_L as shown by

$$C_L = \frac{\text{aircraft weight}}{\text{air density} \times \text{wing area} \times (\text{velocity})^2}$$

And the lower the C_L value the lower the induced drag coefficient C_{D_i}, because induced drag is determined by the equation

$$C_{D_i} = \frac{(C_L)^2}{\pi(AR)}$$

where C_{D_i} = induced drag coefficient

C_L = lift coefficient

AR = wing aspect ratio

So we see that aircraft intended for moderately slow flight, which includes most of our 115 to 150 mph general aviation types, will have undesirable built-in induced drag due to high C_L values unless designed with adequate wing area. While increasing wing area also reduces landing speed and improves climb performance, the wing weight, frontal area, and skin friction drag will be greater; so a design compromise is necessary. Since Chap. 6 covered the effects of aspect ratio and induced drag in great detail under the subject of increasing climb, refer to that chapter if additional information is desired.

REDUCING DRAG I know that many readers are not concerned about drag reduction if their annual flying time is quite low. However, when flying as little as 50 hr a year, a 10 percent reduction in drag will effectively provide 1 or 2 hr free flying. For example, assuming flight over the same distance, the saving possible from a 10 percent drag reduction for an airplane using 9 gal/hr of 75-cent fuel will be

Original annual cost = 50 hr × 9 gal/hr × 0.75 = $337.50
10% drag reduction = 50 hr × (0.9 × 9 gal/hr) × 0.75 =   303.75
Annual saving in fuel alone = $ 33.75

Everyone would enjoy flying faster for the same price, or being able to reduce power and fly at the former cruising speed with less noise and cost. And as added gains, the ROC increases with a cleaner airplane while higher cruising speed with the same fuel consumption is an obvious safety factor when skirting weather or operating in IFR conditions.

There are a few simple ways to accomplish drag reduction on any airplane—in fact some are associated with routine maintenance. Although it is apparent that little can be done to reduce drag types 1, 7, and 8 since they are built into the design, we have ample opportunity for performance improvement by working on 2 through 6. However, even type 1 flat plate drag can be partially overcome when selecting a new airplane; that sleek design with the somewhat tighter cabin and thinner wings will go faster for the same horsepower. Such is the secret of the successful line of Mooney aircraft.

Keeping an airplane clean and polished is a matter of performance efficiency as well as proper maintenance. The smoother the surface, the lower the total skin friction drag. This advice also applies to keeping wing and tail surface leading edges free of mud, insects, and dents; they disturb the airflow, adding to drag and reducing maximum lift. While the outer panels can't retract in flight to reduce wing area, the entire airplane—or at least the wing and tail surfaces—can be kept clean.

The elimination of interference drag probably offers the maximum opportunity for performance improvement. In landing gear design alone, comparison studies noted in Chap. 6 indicate that adding wheel fairings to tricycle gear aircraft will improve climb performance midway to that of a fully retracted gear.

Taking some pains to fair over or enclose exposed nuts, bolts, drain fittings, and similar small items will not only reduce interference drag but also eliminate some likely sources of early icing. And those strut fairings that no longer fit tightly, are misaligned, or possibly just missing, are another fine source of drag and icing buildup. They should be tightly in place and positioned to align with the airflow direction.

Although the various antennas carried today may not lend themselves to arbitrary placement, particularly if the equipment they serve is to work properly, it is usually possible to locate them where turbulent air flowing off an antenna will not fall directly upon a strut or landing gear attachment fitting. These structural joints are probably producing turbulent flow all by themselves without any additional help from the antenna. For most applications a small whip-type communications antenna will serve very well and with much less drag than a large blade, while the short ball-ended transponder antenna also works well with DME (distance-measuring equipment) equipment and at less drag and cost than the usual blade antenna (as will be discussed in Chap. 13).

The climb and cruise performance of many aircraft may be improved by adding wing root and canopy fairings to reduce flow separation. Although such additions are not for the amateur, proper filleting can reduce drag and improve stall characteristics. If you suspect flow breakdown in some area of your airplane, tuft the region using pieces of dark yarn about 4 in. long. They may be attached at their forward end with masking tape and can be observed in flight from the cockpit of another airplane flying 100 to 150 ft away. If the pieces of yarn oscillate a half-inch or more, or if they reverse to curl forward in flight, some degree of flow breakdown is indicated. If you wish to improve the flow, have a qualified aeronautical engineer recommend appropriate design modifications and approve any flight tests. Professional advice is suggested since correction of flow separation can occasionally have a pronounced effect upon flight stability and handling characteristics.

The reduction of pressure field drag becomes increasingly important as an airplane gets older, presenting worthwhile possibilities for drag improvement through elimination of various leaks and openings. Doors should be carefully reshaped to fit the fuselage contour, and additional rubber seal strip added if necessary. This effort will reduce both drag and cabin noise level. Small access holes are frequently drilled in skins to accept socket wrenches or permit visual inspection; these can be carefully taped over or plugged with a flush plastic cap which can be removed as required for servicing.

Loose cowling panels should be refitted and attachment hardware adjusted or replaced to assure positive, nonleaking contact. The cooling air should enter at the cowl front and exhaust through the designed outlet area, not all along the cowling sides, top, and bottom.

If you have the time and interest, it is possible to continue the list of improvements to include more efficient cooling airflow, sealed control surfaces, etc. Since many of these possibilities require fairly sophisticated engineering support, a basic drag reduction program might be started by simply cleaning and polishing the entire ship, adding wheel fairings, and getting the doors and cowling tight. Performance should be noticeably improved by such attention; with today's fuel prices, any time spent to reduce drag will be amply rewarded—and you will have a better and safer airplane.

Drag may frequently be reduced by simply getting the airplane trimmed for maximum cruising speed at a given power setting. Sometimes referred to as "getting the airplane onto the step," a worthwhile increase in cruise will be realized for most aircraft by first entering a shallow dive and then carefully trimming in level flight to maintain as much of the speed increase as possible. The cruising speed of hull-type amphibians can frequently be increased 10 to 15 mph through proper trim attitude. Of course, this is an extreme gain due to reduction of flow separation around the hull chines and step, but such increases indicate the extent of drag reduction possible through proper attention to cruise trim.

COMPARING DRAG   The methods used to determine drag values are of interest, since drag is so basic to all aircraft performance—to say nothing of the importance of high cruising and top speeds to the sales department. The earliest experimenters determined drag values in a rather random manner with each pioneer using his or her own system of coefficients, making direct comparison difficult at best.

For a time *flat plate area* was an established measurement for drag evaluation. This term referred to a flat shape of sufficient area to produce the same drag as the airplane would have in level flight; the smaller the area the cleaner the airplane. Fortunately, during the early 1930s nondimensional performance coefficients became more common through development of increasingly sophisticated wind tunnel test techniques, and remain the standard for current work.

The sphere was considered a primary reference for determining comparative drag values because it was a precise shape and supposedly would provide the same drag reading, even when tunnel flow was not uniformly parallel to the test section walls. In fact, drag readings from a well-tested reference sphere became the basis for determining test section flow quality in newly constructed wind tunnels.

At the present time, and really for the past 40 years, airplane drag values have been derived and compared on the basis of total wing area and maximum level flight speed. In accordance with aircraft engineering practice, total wing area includes the area between the spanwise projection of the wing leading and trailing edges as shown by Fig. 6-8.

Since we are primarily concerned with drag evaluation and reduction in this chapter, the study of 15 representative domestic and foreign private aircraft presented by Fig. 10-3 is of interest for comparison and design purposes. In addition to wing area and airplane total drag coefficient C_{D_T}, wing and power loadings as well as maximum speeds and gross weights have been included.

As noted on Fig. 6-10, the combined value of wing and power loadings W_T should be kept below 28 for acceptable landplane performance. When designing seaplanes the combined loading should be less than 26 for truly satisfactory water operation, but is usually nearer 30 due to greater structural weight of the hull or floats plus the useful load that must be carried to justify the power required. As a matter of interest, data for the popular Piper PA 28-140 Cherokee quite closely approximates the average values for all 15 aircraft of Fig. 10-3.

Airplane drag is compared at maximum level flight speed V_{max} because at that operational point powerplant thrust equals airplane total drag. As a result, an equilibrium speed is established that may be readily converted into the airplane nondimensional total drag coefficient based upon wing area.

This relationship is shown by the point B intersection of the power-required and power-available curves of Fig. 10-2, with the corresponding airplane total drag coefficient value C_{D_T} determined by the following formula:

$$C_{D_T} = \frac{(\text{horsepower}) \ (\text{prop efficiency}) \ (146,620)}{(S_w) \ (V_{max_{mph}})^3}$$

where S_w is the wing area and prop efficiency is usually 0.75.

It is also possible to determine the airplane drag D, in pounds by using

$$D = \frac{(\text{horsepower}) \ (\text{prop efficiency}) \ (375)}{V_{max_{mph}}}$$

and then finding the value of C_{D_T} by the nondimensional formula

$$C_{D_T} = \frac{\text{Drag}}{(1.0756) \ (\rho) \ (S_w) \ (V_{max_{mph}})^2}$$

where ρ = sea-level mass density of standard air = 0.002378.

Rather than become too deeply involved in performance calculations, it is sufficient to note here that nondimensional coefficients let us find lift and drag values by using one basic type of formula.

Airplane model	Total drag coefficient C_{D_T}	V_{max} sea level, mph	Wing area S_w, sq ft	Gross wt W, lb	Horse-power hp	Wing loading W/S_w lb/sq ft	Power loading W/hp lb/hp	Total loading W_T
PIK-15 (Finland)	.0383	142	150.7	1684	150	10.9	11.0	21.9
PA-18 Super Cub	.0451	127	178.5	1750	150	10.0	11.6	21.6
M-B-B BO-209-150	.0339	164	110.0	1808	150	16.4	12.1	28.5
Poulet DM-165	.0305	165	132.4	1850	165	14.0	11.2	25.2
SAAB MFI-15B	.0352	155	134.5	1930	160	14.4	12.1	26.5
ZLIN 42	.0556	136	141.5	2028	180	14.3	11.3	25.6
Maule M-4	.0330	147	152.5	2100	145	13.8	14.5	28.3
PA-28-140 Cherokee	.0384	139	160.0	2150	150	13.4	14.3	27.7
Socata Rallye 150	.0566	130	132.4	2160	150	16.3	14.4	30.7
Grumman American Traveler	.0350	150	140.0	2200	150	15.7	14.7	30.4
Beech B19 Sport	.0412	140	146.0	2250	150	15.4	15.0	30.4
Cessna 172	.0353	139	174.0	2300	150	13.2	15.3	28.5
Lark Commander	.0419	138	180.0	2475	180	13.7	13.7	27.4
Aero Commander 112	.0290	165	152.0	2550	180	16.8	14.2	31.0
Wassmer WA-41	.0341	150	172.0	2645	180	15.4	14.7	30.1
Average	.0389	146	150.0	2125	160	14.2	13.3	27.5

FIGURE **10-3**
Comparison data for domestic and foreign general aviation landplanes of similar power

The references at the end of this chapter contain a brief nomenclature list (10.1) and a glossary of formulas (10.2) used to determine aircraft performance, and speed conversion values (10.3), as well as a series of useful design references (10.4). Although this material also applies to other chapters, it is presented here for want of a better location. In addition, these references provide an excellent data source for readers wishing to study performance calculation in greater detail.

FUTURE PERFORMANCE   As a summary statement it may be safely noted that private aircraft perform-ance has not experienced any "great leap forward" over the past 40 years. Although many technical and operational advances have been realized, greatly contributing to flight safety and reliability, our speeds are about where they were in the thirties. During that period of grass airfields, the four-seat, stream-lined fixed gear Percival Vega Gull of Fig. 10-4 cruised at 160 mph with a 200-hp Gipsy six engine, and offered a standard range of 600 mi. Produced in some quantity for the time with 90 aircraft delivered, this design won the 1936 King's Cup Race against aircraft of similar performance.

What has happened to our engineering over the past 40 years? We have slightly improved upon the Vega Gull's performance by retracting the landing gear on the 200-hp Mooney and Piper designs, but only Robin and Wassmer of France currently offer fixed gear aircraft approaching 1936 speeds.

There certainly is a design challenge to be met, and the Popular Flying Asso-ciation of Great Britain is doing something to encourage useful development along these lines, while also adding a bit of sport to weekend rallys. This group has established a combined aircraft and pilot competition scoring system based upon ROC, cruising miles per gallon, takeoff and landing distances, and the aircraft maximum/minimum speed ratio. Aircraft are separated by categories such as single-place, two-seat, four-place, and twin-engine types. Rate of climb is scored at 1 point per 100 ft/min; miles per gallon clocked over a 1-hr period are rated at 1 point per mile; takeoff distance is added to the landing run, with the sum divided by 2 and scored at -1 point per 100 ft; while the speed ratio of V_{max} divided by V_{min} is reduced by 1 and given 10 points per each remaining unit. Such competition not only adds to the pure enjoyment of sport flying, but

FIGURE **10-4**
Percival Vega Gull
Flight International Photo

also improves pilot technique while encouraging some progressive thought and effort toward more advanced private aircraft development.

Expanding this scoring through an example demonstrating superior aircraft performance, let us first consider a four-place single-engine design climbing 800 ft/min at gross, delivering 18 mi/gal, taking off in 800 ft and landing in 700, with a maximum speed of 170 mph and stalling speed of 55 mph. This airplane would rate as follows:

$$\text{Climb: } 800/100 \times 1 = 8.0 \text{ points}$$

$$\text{Miles/gallon: } 18 \times 1 = 18.0$$

$$\text{Takeoff and landing: } \left(\frac{800 + 700}{2 \times 100}\right) \times (-1) = -7.5$$

$$\text{Speed ratio: } \left(\frac{170}{55} - 1\right) \times 10 = 20.9$$

$$\text{Design total} = 39.4 \text{ points}$$

Now suppose that through revised design the climb rate could be increased to 1000 ft/min, the takeoff run reduced to 600 ft, landing distance to 500 ft, and stalling speed dropped to 50 mph—with fuel consumption and top speed unchanged.

$$\text{Climb} = 10.0 \text{ points}$$

$$\text{Miles/gallon} = 18.0$$

$$\text{Takeoff and landing: } \left(\frac{600 + 500}{2 \times 100}\right) \times (-1) = -5.5$$

$$\text{Speed ratio: } \left(\frac{170}{50} - 1\right) \times 10 = 24.0$$

$$\text{Revised design total} = 46.5 \text{ points}$$

Note the weight given to the short-field performance parameters of stalling speed, takeoff run, climb, and landing distance. All of which translate into cross-country utility and flight safety. This is where our future operational challenge lies: *increasing small-field performance without penalizing efficient cruising speed.*

If all active pilot organizations sponsored sport-flying events of this type on a regional and national basis, I'm sure we'd see more interesting and rapid development of private aircraft design.

Superior powerplant installation, wing design, and weight control at current state-of-the-art levels would permit this improvement as shown by Fig. 10-1 summary results. As future private aircraft move toward such utility, we will eventually reachieve 1936 performance levels, but, unfortunately, at 1980s prices.

NOTES AND REFERENCES

10.1 Nomenclature (See 10.2 for related formulas; see Chap. 15 for nomenclature specifically related to seaplane design.)

a.c.	Airfoil section aerodynamic center (approximately the center of lift)
AR	Aspect ratio = (span)²/wing area = b^2/S_w
b	Wingspan, ft
BHP	Engine-rated brake horsepower = hp
c	Wing chord, ft or in.
C_{d_o}	Wing profile drag coefficient
C_D	Drag coefficient
C_{D_i}	Induced drag = $(C_L)^2/\pi \cdot AR$
C_{D_T}	Airplane total drag coefficient
c.g.	Center of gravity
C_L	Wing section lift coefficient
$C_{M_{ac}}$	Wing section moment coefficient
c.p.	Wing section center of pressure (center of lift)
D	Drag, lb
L	Wing lift, lb (same as W for design work)
M_{ac}	Moment about the aerodynamic center
MAC	Mean aerodynamic chord (wing average chord)
ROC	Rate of climb, ft/min
S_{HT}	Horizontal tail area, sq ft
S_{VT}	Vertical tail area, sq ft
S_W	Wing area, sq ft
thp	Thrust horsepower (= engine power × prop. effic.)
V	Velocity
V_{max}	Maximum level flight speed (ft/sec, mph, knots)
V_s	Stalling speed (ft/sec, mph, knots)
W	Weight, lb (same as L for design work)
W_T	The total of wing + power loading = $W/S_W + W/hp$
α	Angle of attack, degrees
β	Propeller blade angle
δ	Surface deflection, degrees

η Propeller efficiency

ρ Mass density of standard air:

$$\rho = 0.002378 \ @ \text{ sea level}$$
$$= 0.002177 \ @ \ \ 3{,}000 \text{ ft}$$
$$= 0.002049 \ @ \ \ 5{,}000 \text{ ft}$$
$$= 0.001928 \ @ \ \ 7{,}000 \text{ ft}$$
$$= 0.001756 \ @ \ 10{,}000 \text{ ft}$$

10.2   Useful performance formulas:

$$C_L = \frac{W}{(\rho/2)\,(S_w)\,(V_{ft/sec})^2} = \frac{391}{(S_w)\,(V_{mph})^2} \qquad \text{at sea level}$$

$$C_{D_T} = \frac{(\text{horsepower})\,\eta\,(146620)}{(S_w)\,(V_{max_{mph}})^3} \qquad \text{at sea level}$$

$$= \frac{\text{Drag}}{(1.0756)\,(\rho)\,(S_w)\,(V_{max_{mph}})^2} \qquad \text{at any altitude}$$

$$D = C_{D_T}(\rho/2)\,(S_w)\,(V_{ft/sec})^2 \qquad \text{lb}$$

$$L = C_L\,(\rho/2)\,(S_w)\,(V_{ft/sec})^2 \qquad \text{lb}$$

$$M_{ac} = C_{M_{ac}}\,(\rho/2)\,(S_w)\,(V_{ft/sec})^2\,c \qquad \text{ft-lb where } c \text{ is wing chord in ft}$$

$$\text{MAC} = \frac{\text{wing area}}{\text{wing span}} \qquad \text{ft (this is an acceptable approximation for basic design work)}$$

$$\text{ROC} = \frac{(\text{thp}_{avail.})\,(33{,}000)}{W} \qquad \text{ft/min}$$

$$\text{thp} = \frac{D \times V_{ft/sec}}{550} = \frac{D \times V_{mph}}{375} = \frac{D \times \text{knots}}{326} \qquad \text{or}$$

$$\text{thp} = (\text{engine-rated power})\,(\text{propeller efficiency})$$

$$V_2 = V_1 \sqrt[3]{\left(\frac{hp_2}{hp_1}\right)\left(\frac{\rho_1}{\rho_2}\right)} \qquad \text{for variations in power with altitude}$$

$$V_{max_{mph}} = \sqrt[3]{\frac{(\text{BHP} \times \eta)\,348.6}{(C_{D_T})\,\rho\,(s_w)}} \qquad \text{where } \eta \text{ is usually near 0.75}$$

$$V_{S_{ft/sec}} = \sqrt{\frac{W}{(C_L)\,(\rho/2)\,(S_w)}}$$

$$V_{S_{mph}} = \sqrt{\frac{391 \times W}{C_L\,(S_w)}}$$

$$V_{s_2} = V_{s_1}\sqrt{\frac{W_2}{W_1}} \qquad \text{for landing speed variation with weight of the same airplane}$$

10.3   Speed Conversion Values
(ft/sec) (0.592) = knots
(mph) (0.868) = knots
(ft/sec) (0.682) = mph

(knots) (1.1516) = mph
(mph) (1.467) = ft/sec
(knots) (1.687) = ft/sec

10.4 References

(a) D. B. Thurston, "Estimating Plane Performance via Comparison Method," *Aviation,* September, 1946, pp. 55–62.

(b) W. Bailey Oswald, *General Formulas and Charts for the Calculation of Airplane Performance.* NACA Technical Report No. 408, 1932.

(c) K. D. Wood, *Technical Aerodynamics.* McGraw-Hill Book Company, New York, 1935 and later editions.

(d) E. P. Warner, *Airplane Design—Performance.* McGraw-Hill Book Company, New York, 1936 and later editions.

(e) Jacobs and Abbott, *Airfoil Section Data Obtained in the NACA Variable-Density Tunnel.* NACA Technical Report No. 669, 1939.

(f) I. H. Abbott and A. E. von Doenhoff, *Theory of Wing Sections—Including a Summary of Airfoil Data.* Dover Publications, Inc., New York, 1959.

(g) F. E. Weick, *Aircraft Propeller Design.* McGraw-Hill Book Company, New York.

11 POWERPLANT CONSIDERATIONS

Although neither the airframes nor the engines of the earliest aircraft showed any degree of reliability, by 1916–1920 the structures of most production airplanes were superior to their engines. In fact, many of the Jennys and similar types used for barnstorming during the 1920s fortunately had such redundant structures that damage could literally be repaired with bailing wire and any type of wood available—or ignored completely until something fell off or collapsed from constant overloading.

We have come a long way since then; our modern engines and propellers are reliable and represent the major mechanical improvement in general aviation over the past 40 years. Although the problems resulting from discontinued 80/87 octane fuel tend to cloud the picture at this moment, engine development is rapidly adjusting to the higher lead content and aromatics used in available fuels. In fact, the 1977 Cessna Skyhawk is equipped with a new model 160-hp Lycoming engine designed for use with present 100 octane fuels. While operational problems persist with older engines designed for lower octane ratings, engine manufacturers have developed flight and maintenance procedures for these engines to keep undesirable valve and plug deposits at a minimum.

Basically, the powerplant installation consists of the engine, propeller, and fuel and oil systems. However, to meet today's government requirements we must add a fifth consideration: reduction of external noise levels; and this seriously affects engine and propeller design.

Turbine engines will not be covered in this book even though they enjoy a deserved reputation for reliability, plus up to 3000 hr time between overhaul (TBO). Present technology does not permit production of even the smallest FAA certified turbines at a unit price comparable with the total cost of an equipped four-place airplane, while fuel consumption tends to remain high; both these shortcomings preclude turbine use by the average private pilot. Until composite ceramics or similar materials can be used to cast or form turbine rotors and housings it is unlikely that this type of engine can be priced to power standard, high volume production private aircraft. Obviously, the light weight, reliability, and smooth operation offered by the turbine make it a most desirable power source once price and fuel consumption begin to approach reciprocating engine levels. And at the rate our current production reciprocating engines are increasing in price this may not be too distant. Quite likely new materials, resulting price changes, and improved fuel consumption will make turbine power competitive with reciprocating engines by the 1990 time period. In the meantime, let us look at what we have to work with today.

1. RECIPROCATING ENGINES

Required power is determined during the early stages of preliminary design, whether for a completely new airplane or a simple modification of an older model. Due to the extremely high investment necessary to obtain FAA approval for new engines, even when an existing design is being uprated or stretched, certified engines are not available in desirably small increments running from 60- to 300-rated horsepower. Nor are the rotational speeds as low as necessary for reduced cabin and external noise levels. Nonetheless, the designer must work with available engines, while the engine manufacturers and aircraft companies must anticipate power ranges likely to be needed a few years hence. Fortunately, through such communication between the major small aircraft and engine manufacturers, suitable engines are usually developed and certified by the time they are needed for production aircraft.

Commercial transport manufacturers traditionally circumvent new engine problems by selecting an engine backed by military service. We have recently seen a departure from precedent, however, with the latest wide-body transports all developed around new engines. The resulting service problems have been minimal, principally because proven military design features were used on each engine selected.

The worst situation for a designer to face is the combination of an experimental engine installed in an experimental airplane. Military aircraft development is constantly exposed to this costly and frustrating problem because both airframe and engine are joined at the limits of existing technology. Although this condition may be necessary to produce an advanced military vehicle, it must be avoided in private aircraft development to assure product reliability and reduce engineering costs.

Nothing can add to development and flight test expenses as fast as an unproven engine that doesn't cool or will not deliver rated power. Final flight work cannot be started until proper engine performance is realized, and only then can engineering begin tests necessary to develop acceptable handling characteristics and meet FAA certification requirements.

While it is most frustrating to face continuing development delays from experimental engine problems, an even worse situation is to experience repeated service difficulties from a newly certified engine installed in production airframes. The excellent Continental Tiara engines suffered from an undeserved reputation arising from early manufacturing difficulties. This problem was amplified at the time because a few problem engines were installed in recently approved Piper Brave Ag planes being placed in service across the country. If these engines had been proven in an existing and well-known design before being used on a new airplane, any engine problems would have been considered more routine in nature. The aircraft designer should be constantly aware of this potential problem when selecting an engine for a new airplane—and never consider installing experimental engines in aircraft scheduled for production within the next couple of years.

With this overriding consideration, the few approved engines available in the desired power range must be carefully reviewed when finalizing any new airplane design. Such items as rpm, fuel consumption, engine weight, accessory provisions and access, type of engine mount required, and engine reputation with pilots and fixed-base operators all tend to further restrict engine selection within a very narrow field of candidates. With today's pressures upon fuel economy and operating efficiency, the designer must work to reduce airframe weight and drag in order to obtain maximum performance at minimum power. Cost is also paramount, since both engine price and airframe cost increase rapidly with power and weight, offering further incentives for efficient design. While you may not have realized all this effort was necessary to produce that airplane you fly, these design options have been carefully and successfully analysed for every major airplane type in production today.

The standard small aircraft engines presently available in the Western world are of 4- or 6-cylinder, horizontally opposed, reciprocating, air-cooled design. Known as "flat engines," because their principal dimensions lie in the horizontal plane rather than being vertical or circular (radial), various power and fuel system models are manufactured, principally by Continental and Lycoming. Rolls-Royce produces a few Continental designs and modifications under license in England, and Franklin engine rights were recently purchased by Pezetel for manufacture in Poland. Hopefully, Pezetel will make the 2-cylinder 60-hp and the 4-cylinder 125-hp models available to all, since these engines would be useful for powered sailplane and efficient private aircraft designs.

As an approximation, flat engines with normal carburetion deliver about 0.50 hp/cu in. of piston displacement. That is, a Continental 0-200 will be rated at 100 hp, a Lycoming 0-360 at 180 hp, etc.

Engine accessory provisions include fuel pumps, oil flow control and filter systems, self-starters, alternators with electrical output sufficient for the most complex IFR dual avionics installations, vacuum and hydraulic pump drives, governor drives for constant speed propeller control, and on some Continental models, integral oil cooler installations. In addition, the fuel may be delivered to the cylinders through a gravity- or pump-fed carburetor, by pressure fuel injection, or by turbocharged air passing through an inline carburetor. Of course, the fewer the accessory provisions, the cheaper the engine; while installation costs soar from the basic gravity feed to a turbocharged installation which may cost twice as much as a gravity feed system of the same sea-level power rating.

POWER VS. ALTITUDE   Practically all aircraft used for business, pleasure, and sport flying are powered by *sea-level-rated* engines. That means the horsepower value assigned by the engine manufacturer is determined under standard sea-level ambient air conditions of 59°F and 29.92 in. of mercury. Because these engines are not supercharged they are also frequently referred to as *normally aspirated* engines, which is to say they breathe normally and without outside assistance.

Referring to Fig. 11-1 we see that all normally aspirated engines, fuel-injected or not, decrease in power with altitude. However, note how the turbocharged TO-360 holds its sea-level power right up to 10,000 ft. With larger supercharge capacity, rated power may be carried to higher altitudes exceeding 15,000 ft. Since oxygen will be required for sustained flight above 12,500 ft, and since most private flying is at altitudes below 10,000 ft, the ability to carry full power to 10,000 ft is adequate for most flights and extremely helpful in maintaining a high ROC to cruise altitude. Further, the ability to develop full power in the less dense air of such altitudes means faster cruising speeds plus increased operational flexibility in marginal weather and IFR conditions. But at a penalty.

The penalty is: increased empty weight compared with an engine installation of similar sea-level power; greater initial and maintenance costs; and a higher level of flight technique. Note that the 210-hp turbocharged TO-360 weighs 22 lb more than the fuel-injected 200-hp sea-level-rated IO-360. It also costs over 60 percent more. In addition, the TO-360 will have a heavier and more complex exhaust system than the IO-360, as well as more sophisticated engine controls and control procedures requiring pilot attention.

As an alternative, the designer might consider installing a more powerful sea-level-rated engine and restricting full power output until reaching some

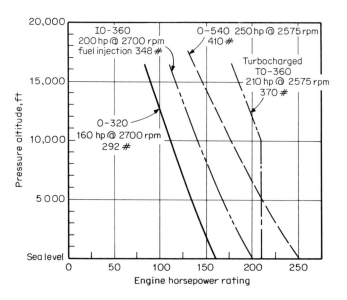

FIGURE 11-1
Power variation with altitude
for some production engines
using 100 octane fuel

designated altitude. In effect, reserve cylinder capacity provides the engine with its own source of altitude performance. Figure 11-1 shows what would happen with an 0-540 engine rated at 250 hp at sea level. This installation could be limited to 200 hp at sea level by regulating manifold pressure through throttle control. But at 6500-ft pressure altitude full throttle output would only be 200 hp, and this drops off rapidly to 177 hp at 10,000 ft. In addition the basic engine weighs 40 lb more than the turbocharged T0-360, so the empty weight will be greater, resulting in a climb penalty if the same useful load is carried. This is not a very effective trade-off, although the 250-hp 0-540 costs only two-thirds as much as the 210-hp turbocharged T0-360. So we see that when altitude power is required, turbocharging is a more efficient design solution than excess sea-level power.

Turbocharging provides some other important although not so obvious operating advantages. Combustion air intake system losses normally waste 5 to 7 percent of sea-level-rated power; by the time this air has entered the duct inlet, gone through the filter, and passed around a few duct elbows and bends, the sea-level pressure—even with some ram effect from flight speed—will have dropped 1.50 to 2 in. of mercury. That means intake air pressure at the carburetor inlet will be around 27.50 to 28.00 in. of mercury instead of the sea-level standard 29.92. Because turbocharging compresses the intake air, full sea-level power can be maintained with turbo operation. This is an important gain providing superior sea-level takeoff and climb performance, thanks to turbocharging.

In addition, the turbocharged engine results in quieter operation both inside and outside the cabin, a most desirable design feature reducing pilot fatigue on long cross-country flights while also making flying more acceptable to today's noise conscious public. Since we will see increasing use of turbocharged engines in the years ahead, a brief description of turbocharger operation follows.

Referring to Fig. 11-2, which is a greatly simplified diagram of turbocharged engine flow paths, first note that combustion air enters the intake duct just as it would for sea-level carburetor or fuel injector equipped engines. The air then goes through an approved filter, past an alternate air inlet door which may be manually or automatically operated or both, and finally reaches the turbocharger compressor inlet. Incidentally, combustion air is usually referred to as *induction air,* with the delivery ducting and various related parts known as the *induction system.*

Once the induction air reaches the compressor stage of the turbo system, any similarity to sea-level engine induction systems stops in a hurry. In principle, engine exhaust gas drives a turbine which is coupled to a compressor. This compresser in turn takes the induction air and increases its pressure before metered fuel is added through a carburetor or injector system. The exhaust gas goes overboard after doing the work of compressing the combustion air. Of course, the fuel and air mixture burns in the cylinder, expanding to provide power through pressure exerted upon the pistons, and in turn becomes exhaust gas working to compress the next air charge. This is why the turbosupercharger is so effective; it is a sort of perpetual motion machine. Engine exhaust compresses the inlet air to increase power, which in turn increases the exhaust flow, which in turn increases the amount of compression possible, etc. However, at some point the maximum structural design limits of the engine will be reached, preventing further increases in intake manifold pressure.

Excessive pressure buildup is controlled by limiting turbine power through waste gate control. As noted on Fig. 11-2, the waste gate is an overboard exhaust port bypassing the turbine. If the exhaust is allowed to flow overboard before it reaches the turbine, no power will be available to compress the intake air above ambient limits. As the waste gate is gradually closed with increasing altitude, more and more power becomes available to compress the less dense altitude air to desirable sea-level limits (or even slightly greater manifold pressures if the engine is designed for such operation).

Waste gate position may be manually controlled by observing manifold pressure limits, or may be automatically managed by a fairly complex system of barometric and oil pressure controls. The manual system is certainly the more simple, with some degree of pilot oversight taken care of by a built-in pressure-relief valve which operates at a few inches above desired limit pressure. And when it goes off, any pilot will remain alert for quite a period.

Manual control of turbocharger power is quite common on our private aircraft, and a number of production powerplant installations are using this system. Since manifold pressures must be noted when operating higher-performance aircraft equipped with constant speed propellers, little more effort is required to monitor and adjust turbine operation with altitude to maintain desired manifold pressure levels.

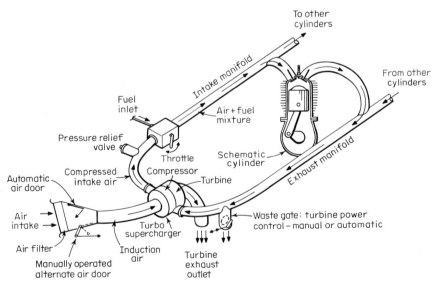

FIGURE **11-2**
Schematic of exhaust gas driven turbosupercharger with in-line carburetor. Fuel injection may also be used with metered fuel delivered directly into the intake manifolds.

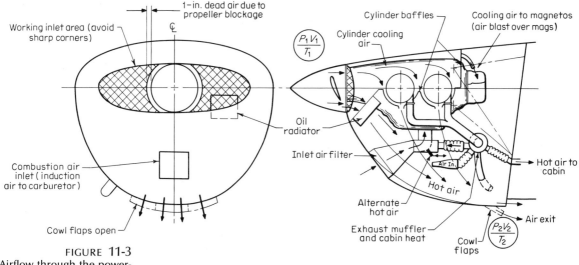

1-in. dead air due to
propeller blockage

Working inlet area (avoid
sharp corners)

Cooling air to magnetos
(air blast over mags)

Cylinder baffles

Cylinder cooling
air

$\frac{P_1 V_1}{T_1}$

Combustion air
inlet (induction
air to carburetor)

Oil
radiator

Inlet air filter

Hot air to
cabin

Air In.

Hot air

Cowl flaps open

Alternate
hot air

Exhaust muffler
and cabin heat

Hot air

Cowl
flaps

Air exit

$\frac{P_2 V_2}{T_2}$

FIGURE **11-3**
Airflow through the powerplant installation

Older and larger engines such as the Pratt & Whitney R-985, R-1340, and up had induction system impellers driven by the engine. These small-diameter compressors operated at very high speeds to increase combustion mixture pressures above sea level, and so provided a degree of built-in altitude performance. When really high altitude power was desired, as for World War II fighter operation, the limits of the engine-driven impellers were extended by combining them with large turbochargers and intercooler radiators necessary to cool the huge amounts of compressed air. As a result, when the outer cowling was removed another solid wall appeared—but this time it was ducting, controls, supports, and baffles.

Fortunately, our small aircraft do not require such complex installations. The advantages offered by the separately mounted turbosupercharger in providing full power at sea level, power at altitude, and quieter exhaust hue to the removal of energy during the compression work cycle combine to make this accessory desirable for high-performance designs. As increasing numbers of production aircraft include turbocharging, the installed price will gradually lower to more acceptable limits; so we should **expect** to see turbopower become more commonplace over the next 10 years. In fact, to simplify powerplant installations as well as reduce engine cost and weight, aircraft manufacturers may be expected to demand that turbocharger systems be built into higher-powered private aircraft engines.

**COOLING
REQUIREMENTS**

Although the cowling should fit tightly to reduce air leakage and should also be of streamlined and appealing design, it must have openings to supply the engine with cooling and combustion air as shown in Fig. 11-3. The cylinder heads and bases, lubricating oil, alternator, and magnetos must receive adequate quantities of cooling air to function properly. FAA test procedures require calibrated documentation of all these items during flight test, with corrections applied to assure cooling under 100°F hot day conditions. If the cowling openings are too large, air will spill over and create turbulent flow, adding to airplane drag. If the openings are too small, the cylinders and oil will soon overheat. As a result, if the airflow volume ratio between climb and cruise flight is very large, that is, considerably more air is required for climb cooling than for cruise power conditions, cowl flaps should be used. They improve cooling during climb by reducing cowling outlet pressure, which in turn drags more air over the cylinders; the cowl flaps are then closed to reduce cooling drag during cruising flight.

Reciprocating engines are really cooled by a combination of air, oil, and fuel. As we all know, air flowing over cylinders or through a radiator carries away heat. The oil radiator needs cooling air because the lubricating oil has picked up heat during its trip through the engine and must be cooled below 250°F to keep from breaking down into a useless, nonlubricating fluid.

The use of excess fuel is a classic although extremely inefficient way to reduce cylinder temperatures. By enriching the carburetor fuel/air ratio jets beyond optimum, unburned fuel remains in the cylinder to cool the valves and heads after the combustion cycle has been completed. Unfortunately, this approach results in uneconomical fuel consumption plus high carbon monoxide levels in the exhaust. In order to properly define the correct fuel mixture ratio for each new airplane design, the engine manufacturer must specify the carburetor setting used for FAA certification flight tests; for all subsequent production aircraft, FAA specifications require the use of engines with an identical carburetor setting.

Engine cooling is also affected by manifold distribution, carburetor inlet airbox design, and many other small but disturbing factors. While whole books have been written on the subject, I believe a simple cooling airflow example will serve our purposes. Engine installation specifications issued by the engine manufacturer list the volume of cooling airflow required over the cylinders, the permissible cylinder head and base temperatures, and the maximum oil temperature allowed. In order to protect their product liability exposure and engine warranty, the engine manufacturer as well as the FAA must approve the powerplant installation; so the cooling requirements must be well met and with some margin to spare.

Referring to Fig. 11-3, let us consider the installation of an engine requiring 2700 cu ft/min of airflow over the cylinders at full power. The critical condition has been determined as a 70-mph climb speed at full power in 100°F hot day conditions. A 4 × 5 in. oil radiator of 20 sq in. rectangular flow area has been selected, with the radiator cooling air supplied from the cylinder air inlet opening. The following calculations provide flow areas of sufficient accuracy for the initial design approximation.

1. *Powerplant Cooling Air Inlet Area*

a. Cooling air required = 2700 cu ft/min = 2700/60 = 45 cu ft/sec at standard conditions

b. At 70-mph climb, flow velocity = 70 × 88/60 = 102.7 ft/sec (where 88 = ft/sec for 60 mph, to convert mph to ft/sec)

c. Assuming a 70 percent efficient cowling inlet design which results in a 0.7 cowling entrance airflow coefficient, the cylinder cooling air inlet area will be

$$\frac{45 \text{ cu ft/sec}}{102.7 \text{ ft/sec} \times 0.7 \text{ entrance coef.}} = 0.626 \text{ sq ft}$$

d. Oil cooler air inlet area $= \dfrac{20 \text{ sq in.}}{0.7 \text{ coef.} \times 144} = 0.198 \text{ sq ft}$

Total inlet area at standard conditions   = 0.824 sq ft

e. To correct for a 100°F day, the air density ratio must be considered

Standard density ρ = 0.002378
100°F density ρ' = 0.00221 (from density charts)

Inlet area for 100°F operation $= \dfrac{0.824 \times 0.002378}{0.00221} = 0.887 \text{ sq ft}$

0.887 sq ft = 128 sq in. of total cooling air inlet area

This area does not include combustion air volume or magneto and alternator cooling air. Note that the two density values given for sea-level standard and 100°F days are the only ones required for most private aircraft powerplant design calculations.

To allow for inlet flow losses occurring around a propeller hub or spinner, the opening on each side of the cowling should be increased by starting the inlet area measurements 1 in. outboard of the spinner or hub diameter. As a further thought, less dense high-altitude air cannot carry away as much heat as that at sea level; so for high-altitude work additional cooling airflow may be required.

2. *Powerplant Cooling Air Exit Area*

a. The outlet area can be based upon the inlet area plus the expansion of cooling air passing over the hot cylinders and oil radiator. This calculation follows the simple gas law

$$\frac{P_1 V_1}{T_1} = \frac{P_2 V_2}{T_2}$$

where P_1 = inlet pressure, lb/sq in.
      P_2 = outlet pressure, lb/sq in.

V_1 = inlet volume of air (or inlet area, for the same inlet and outlet flow velocities)

V_2 = outlet volume of air (or outlet area, for the same outlet and inlet flow velocities)

T_1 = inlet temperature, °F absolute (degrees Rankin)

= ambient temperature °F + 460°F

T_2 = outlet temperature, °F absolute

= outlet temperature °F + 460°F

as shown on Fig. 11-3.

b. Assuming for our example an inlet pressure of 14.6 psi, an outlet pressure of 14.1 psi, and a cooling air temperature increase of 100°F, for a 100°F hot day the exit area becomes

P_1 = 14.6;

P_2 = 14.1;

T_1 = 100 + 460 = 560°F;

T_2 = 100 + 100 + 460 = 660°F

$$V_2 = \frac{T_2 P_1 V_1}{T_1 P_2} = V_1 \left(\frac{T_2 P_1}{T_1 P_2} \right) = V_1 \left(\frac{660 \times 14.6}{560 \times 14.1} \right)$$

= 1.22V_1 (based upon the same inlet and outlet flow velocities as noted above)

c. With the basic entrance area V_1 calculated as 128 sq in. in climb, the corresponding outlet area V_2 = 1.22 × 128 = 156 sq in. during climb. Another set of calculations is required for cruising flight, with the difference in outlet areas ideally provided by manually controlled cowl flaps to reduce cruising drag.

Although this is not an exact series of calculations, it is close enough for preliminary design work and for modifying your powerplant installation to accept a larger engine or a change in operating conditions. Remembering, of course, that any such modifications would be considered of major nature and must be substantiated by cylinder head temperature surveys taken in flight.

Air for combustion and accessory cooling may be picked up and directed from a NASA submerged air intake as shown in Fig. 11-4, which has been taken from NASA data. Based upon drag studies and tests conducted by NACA and NASA in the mid-forties and fifties, this inlet has extremely low drag and lends itself quite readily to cowling design. Once the required air inlet area has been determined at the recommended width to height ratio of 3 to 5, the duct shape follows the simple formula and design of Fig. 11-4.

ALTERNATE HOT AIR   Some controlled hot as well as cold air is essential to safe powerplant operation. In order to prevent carburetor icing or to get rid of icing when it occurs, a source of hot induction air is required for both carburetor and fuel-injected engines. As shown on Fig. 11-3, alternate hot air is drawn from the exhaust

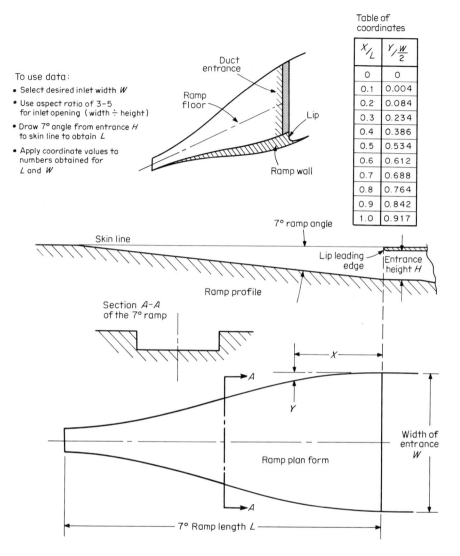

X/L	$Y/\frac{W}{2}$
0	0
0.1	0.004
0.2	0.084
0.3	0.234
0.4	0.386
0.5	0.534
0.6	0.612
0.7	0.688
0.8	0.764
0.9	0.842
1.0	0.917

Table of coordinates

To use data:
- Select desired inlet width W
- Use aspect ratio of 3–5 for inlet opening (width ÷ height)
- Draw 7° angle from entrance H to skin line to obtain L
- Apply coordinate values to numbers obtained for L and W

FIGURE **11-4**
NASA submerged air inlet permits a minimum drag, smooth cowling design.

heater muff directly into the carburetor when the hot-air control is moved to the "on" position. This control is usually grouped on the instrument panel with the other powerplant controls; its operation moves the carburetor airbox hot-air door to shut off the normal cold-air supply and admit hot air instead.

When alternate hot air is applied, the carburetor inlet air temperature rises sharply; in fact, FAA regulations require flight demonstration of a 90°F carburetor air temperature rise for sea-level engines. This test must be performed in 30°F outside air free of visible moisture and with the engine operating at 75 percent power. Since hot inlet air has less density than cold air, making the engine run excessively rich, application of heat causes some loss of engine power. The combination of this power loss plus flow losses through the alternate hot-air induction system combine to make the rpm drop off, as will be noted during engine runup heat checks. In order to keep the heater muff and hot-air delivery hoses from overheating, a small outlet hole is located in the airbox near the hot-air door to keep a small volume of heated air flowing through this system. Whenever possible, it is advisable to inspect the airbox as part of the preflight check just to be sure this opening remains unblocked.

The turbocharged engine does not need an alternate hot-air source since intake air is well heated during its passage through the compressor section of the turbine. Even when the turbine is just idling over, the housing will provide sufficient heat rise to discourage icing. However, an alternate air source is still required because the inlet air filter could ice over; in this case, air may be automatically or manually selected from inside the cowling as shown on Fig. 11-2.

SIMPLIFYING OPERATION

The wet sump oil system with engine oil located right below the cylinders has greatly simplified powerplant installations in small aircraft. The separate oil tank with connecting hoses was cumbersome, heavy, and expensive, although it is still required for aerobatic aircraft if the engine is to keep running during sustained inverted flight.

There are a few other features our powerplant systems could incorporate to improve flight safety while reducing opportunities for "pilot error."

A truly single-tank fuel system with a 5-gal or similar reserve tank, Volkswagen style, would be one such improvement. Although it is most difficult to get two or more tanks sufficiently interconnected to permit equal and uniform flow from each one, I'm sure that properly sized outlet and vent lines plus a feed sump could be developed to permit such simplified fuel control. Of course, we again find a trade-off here in that puncture of one tank or rupture of one of the outlet lines would mean the loss of all fuel. This possibility could be readily overcome in a dual-tank system by using a three-way valve to select tanks "one" or "two" independently, as well as "both" tanks. The difference between such a system and most existing procedures being that selection of "both" tanks would be standard, with tanks "one" or "two" turned on only when separate fuel quantity gauges or visual inspection indicated something was wrong.

In these days of increased fuel economy a simple, cheap, and reliable automatic mixture control would be most useful as well as ensuring maximum duration for all cross-country flights. Such an improvement could be barometric pressure and/or exhaust gas temperature controlled, with manual override retained for pilot operation if necessary.

Another safety device properly receiving increased attention is a spring-loaded suck-in door located downstream from the induction air filter. This door opens automatically to provide induction air if the inlet air filter becomes iced over or otherwise blocked. Note that this feature will not prevent or remove carburetor icing, so a source of alternate hot air is still very necessary—and must be provided for both carburetor and fuel-injected engines.

Returning to the cabin, some standardization of the location, size, knob contours, and knob colors for powerplant controls would be most helpful. I personally prefer the quadrant arrangement over push-pull controls since they seem much easier to identify and more natural in operation, viz. the Cherokee installation. However, hiding the tachometer or manifold pressure gauge behind a control wheel does not contribute to reducing pilot error. In that regard powerplant instruments should be standardized in size, calibration readout, and panel location relative to the powerplant and flight controls. As the tendency increases for pilots to fly more than one type of (rental) airplane, the instrument and powerplant control differences are becoming more critical. This is a subject the FAA has somehow overlooked, and, it is hoped, the General Aviation Manufacturers Association (GAMA) will initiate suitable standards before the government feels obligated to do so.

The following color identification is recommended as a start toward powerplant controls standardization: throttle, black; mixture, red; and propeller pitch, blue. Hopefully, industry will follow up with standardization of the other items mentioned.

As a final note, oil radiators mounted directly on the engine crankcase eliminate the danger of broken oil lines and possible engine compartment fire. Although optional radiators of different cooling capacities would be required to cover different aircraft operating speeds, the simplified installation as well as the safety advantages make this small feature quite worthwhile, as Continental Motors has recognized by including integral oil coolers on some of their engine models.

2. PROPELLER DESIGN

There is really nothing mysterious about propeller design or selection, with the possible exception of deciphering the numbering system all propeller manufacturers seem to assign to their products.

The first propellers were of paddlelike blade shape, made of wood, and of *fixed pitch* design. That is, *the blade angle could not be changed in flight.* One notable exception was a variable pitch, cambered blade, aluminum alloy propeller developed in Italy by Crocco and Ricaldoni in 1907 and used to power a hydrofoil-supported boat at speeds exceeding 50 mph. This seems to have been an unexplained and temporary advance in propeller development, a real uptick in the progress curve, since little serious production effort was directed toward variable pitch propellers until propeller design theory and aircraft speeds made this improvement both possible and mandatory during the 1920s.

Today most of our lower-performance aircraft still have fixed pitch propellers although they are now made of forged aluminum alloy instead of wood, while the higher-performance retractables have constant speed props. Why this difference?

Referring to Fig. 11-5, we find the relative thrust of four propellers plotted against airplane speed. Although these curves are of approximate shape for

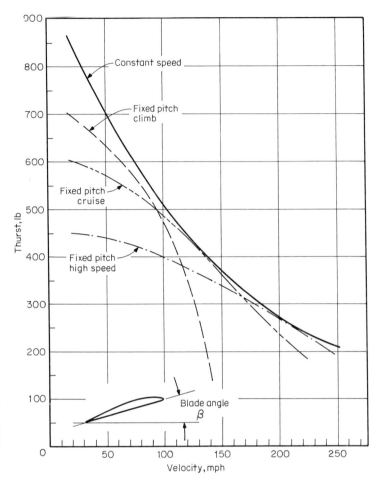

FIGURE **11-5**
Propeller thrust vs. velocity,
constant speed and fixed pitch
types

comparison purposes, it is readily apparent that a *constant speed propeller* encompasses the maximum thrust capabilities of the three fixed pitch designs because it can vary blade angle with airspeed to maintain selected rpm.

The fixed pitch climb prop, used to obtain maximum climb thrust for towing sailplanes or for small seaplane operation, has a low blade angle. This is necessary because best climb is obtained at the low-speed end of the flight envelope and each blade advances forward through the air only a comparatively small distance per revolution; so the inclined plane of propeller advance through the air must be fixed at a small (blade) angle. But even the climb prop cannot be set at as low a blade angle as the constant speed prop, because such a low fixed pitch blade angle would overspeed the engine when the airplane increases from initial takeoff to climb speed. Since the constant speed propeller blade angle will increase with airspeed, the constant speed propeller can be set to a lower blade angle than any fixed pitch type to obtain greater static thrust for improved takeoff performance.

Further, when the fixed pitch climb propeller equipped airplane levels off to cruise, the engine rpm will reach red line long before maximum cruising speed is attained. This is caused by the effect of forward airflow velocity in unloading the blade, and is the result of the low blade angle required for climb performance. And that is why we do not find many climb propellers on private aircraft.

A cruising propeller is the most common fixed pitch type. This blade setting is a compromise between optimum climb performance and operation at or just below rpm red line at level flight maximum speed. As a result static thrust is reduced in comparison with the climb propeller, as shown by Fig. 11-5, although at higher speeds the cruise propeller delivers more thrust because of its greater blade angle. The constant speed propeller has superior takeoff and climb performance compared with the fixed pitch cruising prop, but offers little advantage in the cruising speed regime.

In view of this, why not a two-position propeller instead of a constant speed design? Many two-position models were offered following World War II, but their proper control really required a governor; at which point it became obvious that a constant speed type (which must have a governor) could be offered just as cheaply while delivering optimum performance throughout the entire airplane speed range. And so the successful two-position designs became production constant speed propellers.

As a final fixed pitch comparison, high-speed prop characteristics have been included on Fig. 11-5. The relatively low static and slow speed thrust values result from high blade angle slippage at low forward speeds; the blade angle is so great that it just churns air around instead of efficiently displacing a mass of accelerated air. This explains why the early Cleveland Air Racers ran along the ground such great distances because of the resulting low acceleration obtained from their high-speed props. In fact new designs were frequently just able to take off with the propeller selected for racing performance airspeeds. But once airborne, high blade angle propellers increase in efficiency with airspeed. So we see that by their ability to change blade angle with airspeed, constant speed propellers can include high-performance operation as well as maximum takeoff, climb, and cruise performance. With this capability it is quite obvious why the constant speed propeller is selected for modern higher-performance aircraft.

In another application, although certainly not a high-speed one, seaplane performance is greatly improved by use of a constant speed propeller. Figure 11-6 shows the characteristic increase in static and takeoff thrust compared with a climb propeller. This thrust margin is extremely critical for heavily loaded seaplanes and may make the difference between getting up onto the step and off the water or remaining a low-speed powerboat. Although the constant speed propeller weighs and costs more, and will require some additional maintenance compared with a fixed pitch type, the increased thrust and resulting reduction in takeoff run can be a vital safety feature in rough water operation.

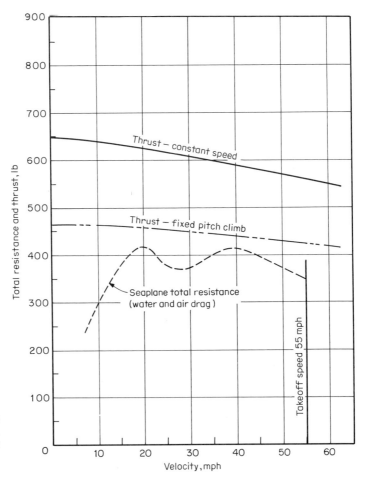

FIGURE **11-6**
Seaplane takeoff drag vs.
propeller thrust

This discussion has shown that constant speed propellers should be used on all aircraft requiring maximum static thrust and cruising performance, regardless of the actual speed range.

Whether the blades are fixed at a selected angle or change pitch to match airplane velocity for constant speed operation, the efficiency of the propeller blade section and planform will have a marked effect upon airplane performance. As presented in Fig. 11-7 for a typical production constant speed propeller, maximum efficiency occurs at cruising and higher speeds. Unfortunately, takeoff and climb occur at the low-speed end where efficiency is very poor. Since blade efficiency is an indication of the amount of engine brake horsepower converted into propulsive thrust, it is important to have propeller efficiency as high as possible throughout the entire speed range. As an example, if the propeller efficiency can be raised 5 percent at an 80-mph climb velocity, the typical 2150-lb airplane ROC will increase from 620 to 675 ft/min; this is about a 10 percent gain in climb for a 5 percent improvement in

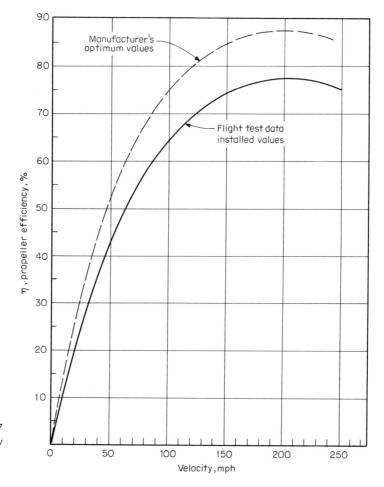

FIGURE **11-7**
Propeller efficiency vs. velocity

propeller efficiency. And propeller operating efficiency is not entirely the responsibility of the propeller manufacturer. You have probably noticed that Fig. 11-7 carries two efficiency curves. The upper one is cheerfully submitted by the propeller designer, while the lower curve shows what we usually end up with when the production installation has been completed—some 10 percent below possible maximum efficiency.

Where did the lost thrust disappear? It was taken a bit at a time by the thick blade root section necessary for structural stiffness and blade attachment, by the hub itself, by the cowling positioned close behind the inner blade section (or by the body turbulence ahead of a pusher installation), by tip flow losses, by cowling inlet flow losses, and by surface irregularities arising from manufacturing tolerances and service erosion. Correcting or improving any of these factors will correspondingly increase propeller efficiency and airplane performance. The spinner is a step in that direction, although some payment is necessary in the form of increased weight and cost. So the airplane designer

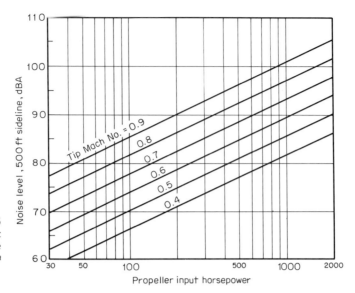

FIGURE **11-8**
Propeller tip speed basic
noise curve
From Hamilton Standard data

can materially improve operating efficiency through cowling refinement, short shaft extensions to locate air inlets further behind the propeller disc, Cessna style, and remote drive shafts that permit fairing the propeller spinner into a small-diameter shaft-support structure.

Tackling the design-improvement problem from another direction, development work by all propeller manufacturers is currently underway to reduce tip losses. This research is intended not only to improve blade efficiency but also to lower propeller noise. Modern engine installations are becoming quieter through use of mufflers and turbochargers to such a degree that propeller noise is becoming increasingly predominant per Fig. 11-8. So EPA (Environmental Protection Agency) noise requirements are indirectly providing additional benefit in the form of decreased tip losses and increased thrust.

Blade efficiency is also being increased all along the span through application of supercritical airfoil sections to blade design. The Advanced Technology Light Twin (ATLIT) development sponsored by NASA has been equipped with Robertson Aircraft/Pacific Propeller designed supercritical propellers as an initial step in developing a new blade concept for general aviation use. A marked improvement in ROC is anticipated and should be realized as shown in our previous example of increased climb resulting from increased propeller efficiency.

Propeller efficiency is also affected by the propeller *activity factor,* an empirical comparative design value that increases with blade width and diameter reduction. Propeller efficiency tends to decrease with the increase in activity factor necessary to deliver more power from a given diameter.

The subject of two- vs. three-bladed propeller frequently arises with an increase in power. In such cases, diameter is an important consideration. For example, study of a recent 250-hp constant speed installation indicated that a three-blade 76-in. diameter propeller was slightly more efficient than a 76-in. diameter two-blade, and much more efficient than a 74-in. diameter two-blade, particularly at climb speed. However, an 80-in. diameter two-bladed design was the most efficient of all throughout the entire flight range, provided such a large diameter could be used.

Increased blade area is required to absorb more power in much the same manner that increased wing area is desirable to support more aircraft weight; the greater the horsepower, the more blade area required for a given propeller diameter. Unfortunately, once a diameter restriction is imposed, an increase in blade area necessary to absorb increased power can result in too high an activity factor value—with an accompanying loss in thrust horsepower. At this point the three-bladed design of lower activity factor becomes desirable since it will be more efficient. And that means it will deliver more thrust horsepower for improved climb and cruising performance.

As a general rule, unless the larger diameter required is limited by tip speed or clearance problems, a two-bladed propeller will be lighter, cost less to purchase and maintain, and be as efficient as a three-bladed propeller capable of delivering the same engine power.

The continuing argument to resolve tractor vs. pusher propeller design advantages is basically concerned with the relative efficiency of these two configurations. Although many sound features can be advanced for either type, the practical maintenance aspects of aircraft operation must favor the tractor design. Any aft-mounted propeller is subject to damage from turbulent flow, materials coming from the cabin, loose hardware left inside the cowling, and debris thrown up by the landing gear. Further, propellers mounted at or above the wing trailing edge tend to reduce wing lift at the higher angles of attack associated with short-field takeoff, as well as during abrupt power corrections in the approach or landing configuration.

The fact that the Cessna 337 aft (pusher) propeller is capable of delivering greater single-engine climb than the forward one would be expected, since the rear constant speed propeller is operating in a regime of reduced velocity airflow. Air is slowed by passing over the fuselage, and, as shown by the comparison curves of Fig. 11-5, thrust increases with decreasing velocity. This condition was particularly apparent on the huge airship *Graf Zeppelin* where the slower speed boundary layer became so deep that the aft engine delivered 15 percent more measured thrust than expected from design calculations.

No propeller review would be complete without a few words concerning noise reduction. Not only is external noise of prime importance from the environmental standpoint, but also cabin noise remains one of the major reasons private flying does not enjoy broad public acceptance. It is not only a social barrier restricting normal conversation, but also extremely fatiguing as well.

Propeller noise is basically a function of tip speed; the nearer tip speed approaches the speed of sound, the greater the noise level as shown by Fig. 11-8.

Although a simple solution might seem to be a direct reduction in rpm, this approach must be accomplished either by gearing the engine or using a powerplant of greater displacement turning at lower speed. Either way, the installation cost and weight both increase for identical power compared with a direct-drive, high-speed engine. In addition, to convert the lower rpm horsepower to thrust will require greater blade area or diameter, or both, with the increased diameter again pushing up the tip speed. Although more blades could be added to absorb the greater torque for a given diameter, a two-bladed propeller becoming three-bladed for example, this solution also imposes an obvious cost and weight penalty as previously noted. In fact, it is really not possible to reduce the noise level of any piece of efficient machinery without paying a penalty in cost, weight, or efficiency—or all three. Nonetheless, noise reduction is a necessary goal for increased acceptance of private aircraft by people both inside the cabin and on the ground.

With this objective, a new effort to reduce propeller noise has been mounted by Hamilton Standard with the Q-Fan concept, while Dowty Rotol is conducting similar research in England with the Ducted Propulsor. Both of these relatively new developments make use of pressure ratio fans operating inside a ducted cowling. While there are some variations in cowling requirements, the major differences between the two designs lie in fan diameter, speed, and pressure ratio. As shown in Fig. 11-9, the Q-Fan is a smaller diameter, higher rpm, and greater pressure ratio unit than the Ducted Propulsor, with preliminary data indicating the Ducted Propulsor is quieter than the Q-Fan at this stage of development. Both of these designs are heavier and more expensive than plain propellers of any type—and considerably more expensive than the standard fixed pitch, forged dural propeller that powers most small aircraft.

When all is analyzed and evaluated, for simplicity, weight, and price, the fixed pitch propeller remains the best average compromise for most low-performance airplanes; and so will be with us as long as reciprocating engines power private aircraft. Readers interested in designing fixed pitch propellers will find detailed procedures, calculations, and charts in Ref. 10.4 (g).

3. FUEL SYSTEM INSTALLATION

A few brief notes about fuel systems should be adequate for our purposes, since actual fuel control is usually well covered in the owner's or airplane flight manual.

Safety should be the primary consideration for fuel system design. And this means no fuel in the fuselage or hull. Fuel should be stored in the wings with the inboard end of each tank at least a foot from the side of the fuselage to keep raw fuel clear of the cabin during any severe accident.

Of course, fuel located in the wings also acts to reduce wing structural loading in flight just the same way wing weight does; both wing structure and fuel have downward acting weights that reduce the flight lift load (acting upward). For this reason, when two or more tanks are located in each wing panel,

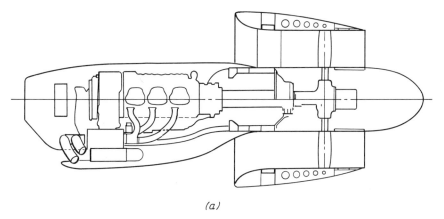

(a)

FIGURE **11-9***a*
Ducted propeller configuration. Hamilton Standard "Q-Fan" installation, left-side view of pylon-mounted piston engine, Q-Fan propulsion pod.

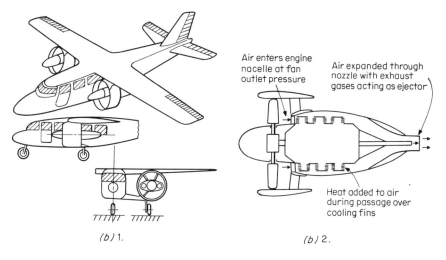

Air enters engine nacelle at fan outlet pressure

Air expanded through nozzle with exhaust gases acting as ejector

Heat added to air during passage over cooling fins

(b) 1.

(b) 2.

FIGURE **11-9***b*
Ducted propeller configuration. Dowty Rotol "Ducted Propulsor" installation. (1) Installation of ducted propulsor. (2) Diagram of thrust recovery from cooling flow.

the inner tank of each side should be emptied first and the outboard tank emptied last; through this procedure the remaining outboard fuel weight will exert the greatest downward moment to reduce wing bending, as discussed in Chap. 12. Note that carrying this to excess by locating all fuel in wing-tip tanks can be dangerous, since in a spin condition weight concentrated so far outboard will tend to wind an airplane into a fast, flat spin.

Fuel tanks should not be located in seaplane floats or down near the step region of amphibian or flying boat hulls. Such float and hull tanks could spray fuel around from striking a submerged object or during a hard landing, and sooner or later every amphibian should be expected to make a wheels-up land landing which could grind the hull down to the step area "built-in-bomb" fuel tank. Again, the wings are an excellent and safer choice for fuel tank location.

As a note of practical caution, mud wasps can become a real problem by building their nests inside fuel tank vent lines. It is possible to start the engine, warm up, and taxi to the active runway without noting any engine or fuel delivery problems. However, if mud wasps have been active, after about 5 min of flight the fuel tank may tend to collapse or the fuel pump quit because of diaphragm failure. And even a thorough preflight frequently will not indicate a clogged vent because these insects may be packing in the mud an inch or two up from the vent outlet. Screened vent fairings would prevent such activity as well as related power failure accidents.

In addition to the convenience of fueling low tanks, being able to look into a tank during preflight to check the actual fuel level is a real safety item as well as a source of personal relief. Unfortunately, many high-wing designs have no provision short of a step ladder for adequate fuel inspection—and this condition deserves further study and design improvement. Conversely, the ability to readily check fuel quantity is a great plus for low-wing and midwing aircraft.

Current FAA design regulations require consideration of lightning protection for all private aircraft. Of course, the best protection is the total avoidance of thunderstorms. But since that is not always possible, grounded-flush-type filler caps and screened tank vent outlets will help. When the system permits, tank vents should be located under the wing near midchord to reduce the likelihood of secondary lightning strikes.

The various types of fuel delivery systems differ in complexity and cost, with a gravity flow carburetor installation being the cheapest. The fuel pump and carburetor combination are next, with the fuel-injected system being even more expensive. Since turbochargers may be used with either carburetor or fuel-injected engines, the fuel-injected turbosupercharger is the most complex and expensive of all fuel delivery systems. This is so not only because of the turbocharger itself, but because of the high-pressure pump and filter required to provide the pressure and clean fuel necessary for fuel injection, as well as the related automatic controls and bypass lines associated with fuel injection systems.

Fortunately, most of our flying does not require the complexity and maintenance cost of either the fuel injector or the turbosupercharger; a designer has little justification for using a fuel-injected 180-hp installation when an equally powerful, lighter, and cheaper 180-hp carburetor-equipped engine is available. And this also becomes a safety consideration since the more complex an installation, the more opportunity for something to fail.

4. NOISE REDUCTION Those of us who have flown for many years, and I'm sure this also applies to everyone really interested in aviation, tend to overlook powerplant noise as something naturally associated with aircraft operation. Unfortunately, occasional passengers or people who are taking a flight to "see what it is like" are usually turned off by the combination of engine and radio noise—even when the radio is as low as it can be and remain audible. By contrast I am frequently driven to admire the captain of an ocean liner or air transport, smug on the quiet bridge with barely perceptible engine vibration to indicate all is well. Without doubt we'd all welcome the same low level of powerplant noise in the cabin.

Noise of any sort is tiring if experienced over an extended period of time. This fact partially explains the popularity of air-conditioned automobiles since they can be driven with all windows closed to exclude road noise, permitting long journeys with minimum fatigue. And I'd gladly yield the quiet of the ocean linear bridge for the sound level of a modern large car, if ways could be found to offer such flight luxury. Although our aircraft save much time by being able to fly in a straight line as well as travel quite a bit faster, I doubt if anyone can fly for 10 hr in any kind of general aviation airplane and not be thoroughly tired by the end of the journey. Yet it is possible to drive in air-conditioned isolation for 10 hr and arrive quite fresh. All this is mentioned to indicate the need to eliminate another potential source of pilot error—fatigue; actually, cabin fatigue, resulting from long cross-country flights under constant engine and radio noise.

Aircraft noise annoyance is not limited to just the passengers, as airport and water-flying restriction suits brought by landowners across the country show only too clearly. Regardless of our personal interests, such restraints are understandable at present noise levels, with the end result being reductions in (1) *aircraft safety* from pilot fatigue and (2) *utility* from prohibited operations.

Fortunately, or unfortunately—depending upon one's viewpoint—we are getting some assistance from the federal government in achieving noise reduction. Although the present and proposed federal requirements are extremely confused, complex, and somewhat contradictory, it appears that *external* sound levels for small propeller-driven aircraft will have to meet Fig. 11-10

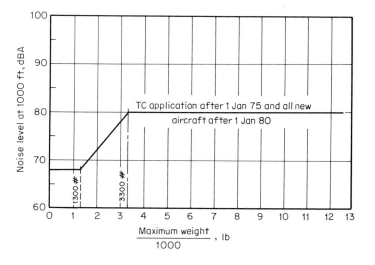

FIGURE **11-10**
Proposed FAA noise standards for propeller-driven small airplanes

decibel (dBA) values by 1977–1980. By the end of this decade some external noise level regulation will be applied to all newly manufactured aircraft except agricultural types, regardless of what date the design was originally type certificated. So we will have quieter flight ready or not; but on the way to this goal you can be sure of meeting some redesign and cost increases which will be briefly reviewed.

SOURCES OF POWERPLANT NOISE

Powerplant noise originates from two principal generators, the engine and the propeller. The propeller emits increasingly high and penetrating frequencies as tip speeds approach the speed of sound as shown by Fig. 11-8. Blade width and propeller solidity also contribute to propeller noise depending upon disc power loading, while tip shape and section contour seem to have some effect as well. Outside of these factors there is little that can be done to reduce normal propeller noise, with tip speed being the most critical consideration.

The engine, however, provides many opportunities for effective noise reduction. As shown by Fig. 11-11, based upon Hamilton Standard research conducted for NASA and corrected to dBA values, energy is delivered from various engine sources besides the crankshaft. It is quite obvious that something should be done to reduce exhaust noise. And when the exhaust sound level is brought low enough, accessory drive, gear reduction, and other crankcase noises will begin to dominate the scene. In fact, only moderate exhaust noise suppression will bring gear and valve noises into focus.

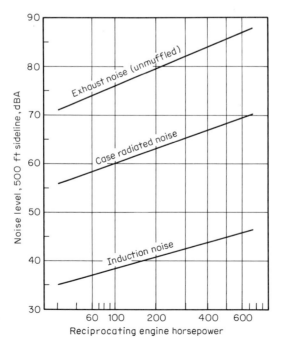

FIGURE 11-11
Engine noise vs. power
(approximate dBA values)

We also have noise produced by accessory items such as the alternator, magnetos, fuel and vacuum pumps, propeller governor, and hydraulic pump that can rival the basic case noise. The entire accessory section and driven accessories are poorly positioned for the average airplane since their normal location is directly ahead of the firewall, permitting all accessory noise to be reflected from the engine right back into the cabin area. The engine manufacturers are finally recognizing this problem and locating more accessory items at the propeller end of the crankcase.

Induction noise is often of very high frequency due to small inlet openings and lead-in ducts, and so tends to influence the entire engine sound spectrum. Fortunately, very little energy is associated with induction flow into the normally aspirated (unsupercharged) engine, and so the flow frequency does not usually vibrate structural panels or other airframe components.

SUPPRESSION OF POWERPLANT NOISE

Certainly the engineering principle of eliminating a problem at the source applies doubly to powerplant noise reduction; the smaller the amount of energy emitted, the less suppression required to lower noise levels to acceptable values.

If we look first at propeller-tip speed noise, an obvious solution to decreasing energy from this source would be a reduction in propeller diameter. However, as the diameter is reduced it may be necessary to add blades in order to absorb engine power, and so the weight and cost of the propeller will both increase as previously noted. Another possibility is the reduction of propeller rpm. Of course, this approach also has power absorption problems, usually solved by increasing propeller diameter—and again climbing back up to higher tip speeds.

Hamilton Standard and Dowty Rotol research engineers are developing another solution to the tip speed problem as discussed under propeller design: a propeller of small diameter and high solidity driven at high rpm, but well below sonic tip speed. This is somewhat similar to a poor man's reciprocating powered turbine or bypass fan jet in principle. Although there is no doubt this design reduces internal and external sound levels, present indication is that such gain is accompanied by increased weight and cost. However, it is still early in the design cycle and additional development may produce a configuration competitive with conventional propeller drive systems but operating at a much lower noise level (see Fig. 11-9).

Current studies are also underway to determine the relationship of tip shape and speed to emitted noise, while other propeller investigators are working on blade width and profile variations intended to reduce propeller noise at increased power output. Something must be done soon, since propeller noise will become increasingly prominent as engines become quieter.

Returning to the engine and its principal sources of noise, mufflers are the obvious answer to the exhaust problem. Although they add weight and cost to the airplane, more and more aircraft are being delivered with mufflers installed as original equipment. Unfortunately, automotive mufflers do not seem to work with aircraft engines, primarily because the automobile is normally operating at a small percentage of total power while the airplane is at or near maximum output. As a result, the combination of auto muffler and aero engine develops excessive back pressure, even with unacceptably large and heavy truck mufflers—so aircraft require their own specially developed, low back pressure mufflers.

When turbocharging is used to obtain power at altitude, exhaust noise is lowered because of the energy removed by driving the turbine and induction air compressor. As noted a few pages back, turbocharging will find increased use on the next generation of private aircraft since it offers faster cruise at altitude plus quieter flight.

Accessory drive and engine case noise would be lowered by decreasing engine rpm. Of course, propeller tip noise would also be lowered, while engine and propeller size and weight would tend to increase for a given power output. Again, an engineering compromise and trade-off of some sort are necessary, including a completely new family of accessories. Improved gearing would also lower accessory and internal drive system noise. Continental has accomplished this with their Tiara series 6-285 and 6-320 engines. Despite a comparatively high engine rpm, the propeller and accessory items are geared down 2 to 1 with a system so designed that the case sound level is noticeably lower than that of more conventional engines of similar power. So it can be done, but time and money are required.

5. EFFICIENT ACCESSORIES   In bringing this chapter to a close, a brief mention of three useful powerplant monitors would seem in order. These are the cylinder head temperature, exhaust gas temperature, and carburetor icing indicators—and they all contribute to operational safety as well as powerplant efficiency.

The cylinder head temperature indicator should be mounted on the hottest head as determined by flight test. Since the airplane manufacturer has demonstrated acceptable engine cooling under adverse conditions during the certification flight test program, one importance of monitoring cylinder head temperature, strangely enough, is to keep the engine from operating at too low a temperature. Low cylinder temperatures mean too much air is flowing through the engine, causing excessive cooling drag and probably inefficient fuel consumption as well.

Low cylinder head temperature may simply indicate the cowl flaps have been left open during cruise, or possibly that the cooling air inlet and/or outlet are too large in area. At any rate, excessive cooling is not good for the engine and should be adjusted to approach more closely cylinder head limits as corrected to 100°F ambient conditions. In addition, cruising speed suffers when cooling drag is too high.

At the other end of the spectrum, a sudden increase in head temperature indicates that something has gone wrong and that power should be reduced at once. Possibly an exhaust stack developed a crack, a valve has gone, the cylinder wall has blown, or an injector has partially clogged—not comforting prospects, but early recognition of trouble provides time to find an acceptable landing spot.

The exhaust gas temperature indicator is of value when fine leaning is necessary to extend cruising range by operating at peak fuel consumption efficiency. Leaning in this fashion should be accompanied by head temperature monitoring to assure that fuel distribution does not change sufficiently to raise any one cylinder much above the average temperatures. By leaning to optimum recommended exhaust gas temperature it is possible to decrease fuel consumption to such a degree that the manufacturer's advertised fuel consumption rates may actually be realized—and this frequently would seem to be quite a feat in itself.

Although carburetor icing can be a problem at any latitude, it is particularly disturbing in northern climates at almost any time of the year. ARP Industries of Huntington, N.Y., has developed a patented indicator that among other duties signals by red dash light when icing conditions develop in the carburetor throat. This warning indicates the need for carburetor heat in sufficient time to prevent icing buildup—certainly an important safety device for powerplant induction systems prone to carburetor icing, which is why this manufacturer has been mentioned.

While many of us have flown for many years without benefit of these aids, their installation could make life a bit more pleasant—and safer as well.

*　　　*　　　*

In conclusion, considering the reliability of current production engines and propellers, the major requirements for an acceptable powerplant installation are

1. A well designed and tested fuel delivery system

2. A properly filtered air induction system incorporating an automatically operated emergency air intake door

3. An adequate source of alternate hot induction air to prevent or remove carburetor icing

All other design considerations reviewed relate to operating efficiency, noise reduction, or pilot acceptance. As such they are important powerplant criteria, but must take a secondary position to the above three primary safety features.

12 AIRFRAME DESIGN

In starting this chapter, it seems only fair to note that a text of this scope cannot possibly cover airframe design in great detail. In view of this, significant criteria and load conditions will be presented only for major components, with particular emphasis placed upon the different methods and materials used for structural design.

Early aircraft were fabricated from bamboo, wood, steel tubing, and wire not only because these materials were readily available but also because high-strength aluminum and steel alloys had not been developed. Although the German Zepplin LZ-1 of 1900 used formed aluminum sheet for the girders and gondola cars, some time passed before the cost and strength of aluminum alloy reached levels necessary to replace wood, mild steel, and fabric as an airframe structural material.

By contrast, practically all current production aircraft are of sheet metal construction. This type of structure has evolved partially because the quality-control standards applied to aluminum alloy sheet, bar, extrusion, and tubing assure repeatability of material thickness and strength. In addition, the load transfer across sheet metal riveted and bolted joints can be calculated very precisely with existing analysis procedures, which helps keep ulcers under control in the stress department. And as a final reason, sheet metal assembly requires less time and less specialized skills than welded steel and fabric construction. Although wood is easily worked, the resulting structure tends to be heavy and subject to deterioration from prolonged exposure to weather.

Unfortunately, sheet metal aircraft manufacture requires a considerable investment in heavy machinery, tooling, and equipment as well as great quantities of generally skilled hand labor. And to magnify the cost problem, the limited volume of aircraft production defies application of known automation techniques. No automobile manufacturer would consider a single run limited to 15,000 units, yet that was the total United States general aviation production for 1976, divided among 10 different companies. Even though three manufacturers produced about 90 percent of this total volume, the largest individual quantities were still comparatively small when the various models are considered.

Since aircraft cost is of basic interest to everyone and is directly related to airframe design, let us follow through the procedure establishing the market price of a new four-place, retractable tricycle gear airplane. In the process we will review the rather abstract production learning curve that is frequently mentioned but never defined when discussing aircraft pricing.

COST VS. QUANTITY   Very early in any cost study, the type of airframe structure used for a new design will be compared with those of prior production models in order to estimate manufacturing labor-hours requirements. For the average low-to-medium-performance sheet metal airplane produced in this country, the parts and assembly time for the one-hundredth unit will total approximately one labor-hour per pound of manufactured airframe weight. Since this weight does not include the engine, propeller, wheels, brakes, tires, and major hardware items, these components are deducted from the basic empty weight to establish the manufactured airframe weight for costing purposes.

For our example, engineering has determined the airplane empty weight to be 1350 lb. The weight of the manufactured airframe then becomes

Basic weight empty		1350 #
Engine weight (equipped)	300 #	
Propeller weight (constant speed)	50 #	
Wheels, brakes, tires and tubes	40 #	
Major hardware items	50 #	− 440 #
		910 #

From this point on, the airframe parts and assembly cost estimates are established by the *learning curve*. This is a logarithmic line plotting aircraft volume vs. manufacturing labor-hours to indicate the probable time required to build each airplane upon a continuing production basis. For our relatively simple new design, production is estimated to move along an 85 percent learning curve as shown by Fig. 12-1, which has been prepared to cover the first 100 aircraft. The 85 percent value refers to a production improvement rate in which each succeeding airplane requires only 85 percent of the time needed to complete the preceding one. If the airframe had been much more complex and estimated at many more labor-hours, production experience might have indicated a more rapid decrease in labor-hours as improved tooling continued to become available; in which case the learning curve might have been established on a steeper 80 percent slope basis, as given in the unit cost table of Fig. 12-1.

Our particular *unit cost curve* has been established by spotting 910 labor-

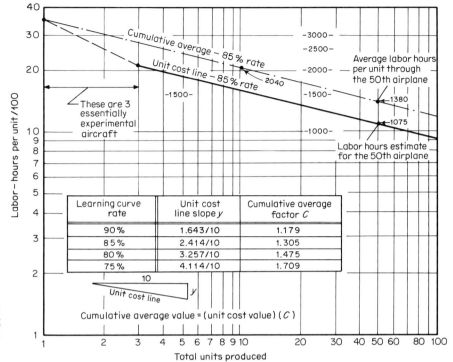

FIGURE **12-1**
A typical production learning
curve and values for plotting
different learning rates.

Learning curve rate	Unit cost line slope y	Cumulative average factor C
90%	1.643/10	1.179
85%	2.414/10	1.305
80%	3.257/10	1.475
75%	4.114/10	1.709

Cumulative average value = (unit cost value) (C)

hours at the one-hundredth airplane and running an 85 percent learning curve through that point. The slopes for the normally used unit cost lines and the factors for the closely related cumulative average cost lines are given on Fig. 12-1 for use with 2×2 or 2×3 cycle logarithmic paper.

Although the actual production hours estimated for any unit will lie on the unit cost line, it is an unfortunate fact of life that all aircraft built up to that point have incurred labor-hours costs indicated by the *cumulative average line*. The cumulative average line value at any production point represents the average number of labor-hours per airplane required to produce aircraft up through that unit. For example, at the tenth airplane the 85 percent cumulative average line indicates that 2040 hours will be the average labor expended per plane for the first 10 units produced; that is, 20,400 labor-hours will be required to produce the first 10 aircraft. Since they are based upon estimates, learning curves should be considered an approximate indication of aircraft production time requirements. But they are the best guide we have to date and do provide historical records for future manufacturing estimates if kept up to date as production develops.

Looking at the fiftieth airplane, we find 1075 labor-hours of unit cost line time scheduled for that unit although we know the company has supported production buildup costs averaging 1380 labor-hours per airplane up through the fiftieth unit, as indicated by the cumulative average line. The difference is 15,250 (305 × 50) labor-hours; and someone must pay for all the time expended in getting production under way and up to the fiftieth airplane. Remember that our pricing estimates were based upon the one-hundredth airplane produced, so even more labor costs will have to be absorbed following the fiftieth unit before the program reaches an assumed break-even point at the one-hundredth airplane.

These extra labor-hours should be anticipated and included in the airplane sales price along with engineering development costs. If this is not done, the capital invested to develop model no. 1 will not be available to develop a subsequent model no. 2. Unfortunately, this condition is not realized by most small engineering-oriented operations, and is one of the primary reasons they drop out of existence with such regularity. Any group considering aircraft production must plan accordingly and obtain sufficient capital to stay alive until sales deliveries can support current costs.

At this point in our cost review, the economics of plant location, taxes, wage rates, invested capital, labor relations, and a number of less important factors enter the equation. Since the manufacturing wage rate including all overhead may vary from $12 to $20 per hour, it could cost anywhere from $10,920 ($12 × 910) to $18,200 ($20 × 910) just to build this airframe. And that is some spread. With raw materials, hardware, engine, propeller, and other purchased parts priced at $10,700 our airplane could cost from $21,620 ($10,920 + $10,700) to $28,900 ($18,200 + $10,700) to produce without including any manufacturing profit, amortization of development and production costs, sales costs, or dealer markup. Assuming a total wage rate of $14 including all overhead charges, the production cost of this airframe at the one-hundredth unit would be:

$$\$14 \times 910\# = \$12,740 \quad \text{assembled airframe cost}$$
$$\underline{10,700} \quad \text{purchased parts cost}$$
$$\text{Production cost} = \$23,440$$

If we assume the design, tooling, and production startup costs totaled $3,000,000 to be amortized over 1000 aircraft, we have another $3000 per airplane to add before any profit or dealer markup enter the picture. If such costs were amortized over only 500 airplanes, the base price would jump a prohibitive $6000 over production cost; so such rapid development write-off is rarely attempted with small private aircraft. But how many second and third-level lightplane manufacturers ever produce even 200 units of a new model? And frequent nonrecovery of invested capital is one reason bankers do not care for the private aircraft industry, and also another reason why undercapitalized new airplane companies go out of business after a relatively short period of operation.

Getting back to the final price for our new four-place design, carrying the introductory costs over 1000 units, we have

Production cost	$23,440
Development amortization	3,000
Subtotal price	$26,440

In order to stay in business, the manufacturer must realize sufficient profit to satisfy the shareholders, provide for plant and equipment expansion and replacement, and have funds for new product development. Also, if this new design is to sell it must be advertised. Experience has shown that advertising and sales expense will run about 10 percent per unit. On this basis our exfactory price becomes

Subtotal price	$26,440
Advertising costs	2,644
	$29,084
Profit at 10 percent pretax	2,908
Price exfactory	$31,992

Dealers need sufficient margin to permit some trading or list price reduction, so their share of the total is usually in the vicinity of 20 to 25 percent of list price, depending upon the airplane price bracket and parent company policy. If we are to adjust the factory price for a dealer discount of 20 percent off list price, we have

$$\text{List price} = \frac{\text{factory price}}{0.80} = \frac{\$31,992}{0.80}$$

List price offered through the dealer = $39,990

It didn't take long to get from the $23,440 production cost to the $39,990 list price, did it? And we haven't included any avionics, IFR instruments, or special equipment options.

These are very reasonable production costs under existing economic conditions and will differ very slightly on a worldwide basis—excepting forced labor economies—as a glance at the Sierra 200, Cardinal RG, Commander 112 series, and Socata ST 10 prices will soon confirm. Although it may seem expedient to subcontract the airframe to a comparatively low labor rate country, by the time necessary materials are imported into that country, the people there finally learn what they are supposed to do, and the airframe is shipped back into this country, the cost will be just as high as with domestic manufacture—but the quality will be much lower. So we just cannot get prices down by using low technology labor; but we can lower costs through a simplified design approach reducing the number of structural parts. The design challenge then lies in simplifying the airframe without unduly penalizing either performance or useful load. Let us see what can be done.

COST VS. DESIGN   Seeing the word cost appear in a subheading for the second time may make you wonder if we are not straying far afield from airframe design. Such is not the case, since airplane cost is directly related to airframe complexity as just shown in the production pricing example.

The lightest airframe for a given set of load conditions will have light skins plus many supporting ribs, stringers (skin stiffeners), and frames. It will also be the most costly structure to tool and produce, and unless well-tooled and carefully built the skins will appear full of wrinkles at rest and in flight. These slight surface variations can be eliminated by increasing skin thickness, while the weight added by heavier skins can be somewhat decreased by increasing rib and stiffener spacing and so reducing the number of airframe parts.

This design philosophy applies to all small aircraft produced in quantity in the United States, and also extends to modern military aircraft where the skin plating may be so thick it is machined to fit the airframe contour. Using wing structure as an example, we find that updated versions of old two-spar, fabric-covered wings may retain both spars even though redesigned for metal covering. The modern trend is toward a single beam wing having a torque tube leading edge and a light rear spar which picks up the wing torque and drag loads. This type of stressed-skin wing construction is shown in Fig. 12-2 and lends itself to integral leading edge fuel tanks, as also used on the similarly designed Cherokee series; a minimum number of detail parts; and a simple three-bolt attachment or similar connection to the fuselage or hull.

A version of this type of wing structure is also used on Rockwell Commander

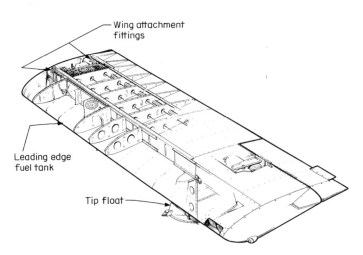

FIGURE **12-2**
Stressed-skin wing construction used on Teal amphibian.
Based upon Flight International drawing

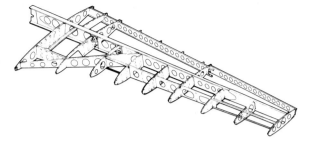

FIGURE **12-3**
The fail-safe wing structure
of the Rockwell Commander,
showing the alternate load path
provided by the inboard leading edge beam.
*Rockwell International-General
Aviation Division Illustration*

aircraft, which have been designed to FAR Part 23, amendment 7. These FAA design specifications require that at cruising speed a wing must be able to support 75 percent of the critical limit load *after failure* of a principal structural element (Ref. 12.1), and that it be able to do so with at least a 15 percent load safety margin. To meet this fail-safe requirement it is essential that efficient alternate load paths be made available, as shown by the diagonal beam located at the leading edge root section of Fig. 12-3. If the main beam root fitting should fail, load would pass along the leading edge beam and be taken out by a leading edge-to-fuselage attachment; all of which adds many parts back into our simplified structure. The wing drag load (acting forward or aft) is usually picked up at the fuselage from the inboard end of a small beam located near the wing trailing edge as shown in Fig. 12-2 and 12-3, and does not require an additional leading edge attachment (which has been added just to satisfy the fail-safe design condition).

The fail-safe requirement of Ref. 12.1 really should not apply to aircraft that may be used only 200 to 500 hr a year, and rarely, if ever, experience a 3 g flight load condition. Analysis of a similar structure for the Teal amphibian indicated a fatigue life expectancy of over 15,000 hr if the structure were loaded to 2 g every minute for the entire 15,000-hr life of the airplane. At an annual use of 300 hr under such turbulent conditions, the airframe would be expected to last for 50 years, showing how ridiculous this FAA transport design requirement becomes when applied to general aviation use. The entire airframe would probably corrode long before fatigue could affect the principal structural members, while the owner would probably also wear out and succumb during the same 50 years.

Despite such unnecessary and costly requirements, a simple wing can be developed around very few parts by designing with stressed skin construction. And the fewer the wing parts to be built and assembled, the fewer the labor-hours required, resulting in a less costly airplane as previously shown.

This design approach obviously applies to the fuselage and tail surfaces as well, explaining the relatively great distances between frames and ribs of most aircraft. But the point to remember is that the lightest structure requires many parts and much labor, while a practical structure of affordable price will have fewer parts and a somewhat greater airframe weight.

FUSELAGE DESIGN   The fuselage should be designed to satisfy two major criteria: (1) protect the passengers in the event of a crash, and (2) efficiently tie together the power-plant, wing, landing gear, and tail surface loads. Of course this must be accomplished with great interior volume for passenger comfort and minimum frontal area and contour drag for maximum performance, another set of challenging compromises for the design team to solve.

Probably the safest and most rugged passenger capsule is provided by the amphibian as a by-product of the hull bottom design load conditions. Since the special requirements of seaplane design will be covered in Chap. 15, we will not pursue this type further at the moment.

Ideally, passengers should be protected by a structure that will withstand a forward loading of 40 g based upon the airframe weight. Metal sailplanes designed to this criteria have saved the lives of pilots who would have been killed by similar crashes in fiber glass fuselages. As an added consideration, if wings are designed to fail forward around the 20 g impact level, the fuselage structure need not absorb wing weight energy at impacts above 20 g, providing a form of inherent load relief that will reduce protective structure weight.

If 40 g protection cannot be realized without paying an extreme weight penalty, some reduced level of crash safety should be provided, extending down to 20 g as a lower limit. Shoulder harnesses that can definitely support a 100-lb upper torso at 40 g should also be provided, and should be anchored into some portion of the structure that can actually pick up and distribute 4000 lb into the fuselage. Reference 12.2 and Chap. 13 discuss seat belt and harness restraint systems at greater length from the equipment installation standpoint, so noting fuselage attachment loads will be our only consideration here.

The reduction of passenger fatalities resulting from off-field landings is receiving increased attention from both the FAA and NASA. This government interest in crash safety is a good indication that all newly certified designs may soon require substantiation of passenger protection during crash conditions. Although structural analysis may prepare the way, it appears that impact demonstrations may be necessary, using an actual airframe structure and dummy passengers suitably instrumented to prove that occupants could survive a crash landing of up to 30 g forward and 15 g vertical loadings—which is why the 40 g forward impact condition has been recommended for fuselage design.

NASA in association with the FAA has been crash testing airframes of modern design under controlled conditions, with details of each impact recorded through multiple instrumentation and camera locations. At contact angles as low as 15° with the ground and at 30-mph impact velocity, critical parts of instrumented passenger dummies have yielded whip loadings approaching 24 g vertically but only about half that magnitude in the horizontal direction. Presumably the horizontal g values will climb rapidly as impact angles and velocities are increased.

It is hoped that this testing will provide useful design criteria for safer structures, seating, and restraint systems, without too many new certification requirements.

Detailed studies of personal and agricultural aircraft crashes based upon the ratio of fatal accidents to total accidents have shown that the least safe personal airplane had 3.46 times the fatality record of the most crashworthy design, while the worst Ag plane had 14.5 times the fatality ratio of the safest model currently available (Ref. 12.3). With a spread of this magnitude, we apparently have some way to go before adequate crash safety is incorporated into fuselage design.

Surprisingly enough, little difference in crash survival could be determined between high- and low-wing aircraft—with the high-wing type actually slightly safer. But this finding could be the result of a trade-off, since most high-speed personal aircraft are of low-wing configuration and a greater percentage of high-speed aircraft accidents are fatal.

As a basic criterion, the safe fuselage design includes a passenger cabin capsule surrounded by energy-absorbing structure capable of supporting the passenger seat belt and shoulder harness restraint systems during impact.

Let us now move on to the fuselage design load conditions. As briefly noted in Chap. 7, the larger the vertical and horizontal tail surfaces, the greater the aft fuselage loading and structural weight. Figure 12-4 is a greatly simplified presentation of this tail loading condition, but it will serve to show how aft fuselage loads are generated. When the horizontal tail is loaded symmetrically as shown, that is, with the same amount of load on each side of the vertical tail, the fuselage will tend to bend down as loaded in our diagram; and the further

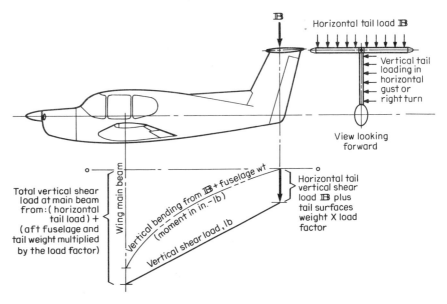

FIGURE 12-4
A simplified diagram of aft fuselage loading from horizontal tail load, aft fuselage weight, and vertical tail weight.

this load travels forward from the tail the greater the bending moment, since the arm distance to the tail load increases (moment = arm times load).

This bending must be resisted by the fuselage aft structure, and the further apart the supporting structural members or longerons can be placed, the lower the resisting load becomes. But as the fuselage is made deeper to reduce bending loads and so lighten the longerons, the skin area increases and becomes heavier. Some design study is usually advisable to review combinations of skin, stringer, and longeron sizes for weight optimization, but may not be particularly critical since passenger accommodation will probably establish the cabin and aft fuselage shape regardless of load requirements.

In addition to vertical bending, our illustration also contains *vertical shear*. Shear of this type is probably best understood as a load acting as though it could be applied at every point along the aft fuselage. That is, the horizontal tail load acts with the same (shearing action) load value all along the aft fuselage right up to the wing beam. For our purposes the shear load is supposedly balanced out at that point by countering shear forces acting downward ahead of the beam, as provided by the powerplant, passengers, and equipment.

Of course more than just the horizontal tail load is acting down, since the weight of the fuselage and tail surfaces also acts down for the condition shown and must be added to the horizontal tail load. The structural weights must be multiplied by the maneuvering load factor to obtain the actual design load, increasing both the bending moment and shear loads for this design condition.

All this may seem complicated at first glance, but really is not if taken one step at a time. You will be pleased to know this load review will not become involved with detailed calculations better left to structural design manuals, but we should be aware of other tail load conditions that serve to compound fuselage analysis.

Returning to Fig. 12-4, note that the view looking forward contains a vertical tail loading that could result from a horizontal gust or maneuvering condition. Assuming the pilot applied up elevator and right rudder at the same time to effect a climbing right turn, we would have the loading shown. Just as the horizontal tail down load caused aft fuselage bending and vertical shear, the vertical tail load shown would result in the application of horizontal bending and shear in addition to the vertical loadings previously discussed.

Although the combined effect of vertical and horizontal tail loadings on the fuselage need not be investigated for so-called conventional tail locations having the horizontal tail positioned on the fuselage, this interaction must be analyzed for the T tail as noted in Chap. 7. The resulting load conditions place the aft fuselage in torsion as well as bending, since the vertical tail center of pressure is located well above the fuselage. Fuselage torsional loading can be further increased by the unsymmetrical horizontal tail load condition required by FAA design specifications. Study of Fig. 12-4 will show how fuselage torsion develops from a partial load on one side of the horizontal tail, full load on the other side, and a loaded vertical tail. Of course, these tail loads can act up or down and left or right, so all possibilities must be investigated.

While we will not go into detail which would become largely repetitive, similar bending occurs in the forward fuselage through application of vertical and horizontal design load factors to the powerplant, with torsion introduced into the engine mount and forward fuselage structure from engine torque. In addition, the nosewheel applies vertical and lateral bending and shear plus torque onto the structure—and can develop all these loadings at the same time during a sideslip nosewheel contact. So land with the nosewheel clear for your own safety and to preserve the forward fuselage structure.

The wing carries bending and shear loads into the fuselage where they must be taken out through strong connections and balanced by inertia forces. This is why a load carry-through beam or set of compression tubes is carried across the fuselage, frequently just where the seats should be located. The wing attachment and fuselage carry-through structure must also be designed for an unsymmetrical wing loading of 30 percent maximum flight load, that is, full lift load applied on one side of the fuselage and 70 percent of that load on the other side (Ref. 12.4). This condition, which will tend to load and rack the fuselage in torsion at the wing attachment points, must be resisted by the fuselage carry-through beam and then transferred into the fuselage structure.

Typical examples of modern fuselage skin and stringer construction are illustrated in Fig. 12-5. Figure 12-5*a* shows the reinforcement required around a door cutout in longeron-and stringer-type structure, while 12-5*b* presents an advanced type of aft fuselage design. Note that this fuselage is assembled in halves joined at the top and bottom centerline. The frames are rolled integral with the skin, with light stiffeners providing skin panel support between frames. Although considerable tooling would be required to make such a simplified assembly, this design approach is all the more remarkable because it was developed and used on the Messerschmitt ME 109 and 110 fighters back in 1935–1938.

A similar structure was used on the two-place Ercoupe designed in this country by Fred Weick in the late thirties and produced in quantity right after World War II. Nothing simpler has appeared on the scene since that time, and really cannot, because the structural configuration of Fig. 12-5*b* contains only the essential parts required to carry the bending, torsional, and shear loads applied by the tail surfaces.

While on the subject of fuselage design, a note or two about windshields and cabin entry would seem useful. In order to reduce aerodynamic noise the windshield should be faired into its supporting structure, with the area around and aft of the windshield designed with smoothly curved contours. Air leaks should be well sealed since they are another source of noise as well as increased drag. Propeller noise can be considerably reduced by a thick windshield, which will also act as its own damper to reduce or eliminate noise

(a)

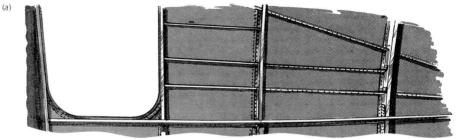

(b)

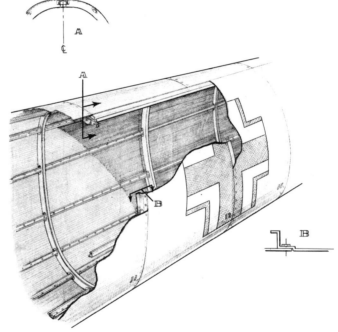

FIGURE **12-5**
Examples of well-designed
fuselage structures. (a) Typical
longeron and stringer construction showing reinforcement around a door opening.
(b) An excellent example of
simplified design. Note that the
frames are formed integrally
with the skin.

originating from thin windshield vibration. Of course, cost and weight penalties accompany heavy glass, but they can be reduced somewhat by eliminating the center post which is not required to support a thick windshield—and
so forward visibility is also improved.

A lightly tinted windshield glass will reduce glare and solar radiation, with such tinting particularly desirable for overhead viewing panels. The tinting must not be too dark, however, because night vision will be reduced to unacceptable levels.

Cabin access seems to be a constant problem with small aircraft. Whether high-wing or low, a tangle of feet, charts, and baggage seems to occur at each end of every flight. Since it is usually older people who can afford to purchase their own airplane and who have the most difficulty entering it thereafter, it seems strange that more attention is not directed toward solving this problem. Probably the Cardinal offers the best entry of any production private airplane design, but all other door-equipped models are quite clumsy.

One solution to the entry problem frequently used in Europe and by Grumman American in this country is the sliding canopy shown in Fig. 12-6. This canopy can be moved back to ventilate the cabin on the ground, offers the safety of 360° visibility, and may be opened in flight for the true feel of sport flying. The only real drawbacks are (1) the need to reduce excess solar radiation through the upper canopy; (2) difficult access to the rear seating if the canopy does not move sufficiently far aft; and (3) the fact that everyone may get wet when the canopy is opened in the rain. Overhead glare can be eliminated by application of heavily tinted plastic liners or the use of darkly tinted glass for the top panels, and proper track design and attention to fuselage lines will

FIGURE 12-6
Sliding canopy on Grumman
American designs
Grumman American Aviation Photo

permit canopy movement sufficient to open the entire cabin if necessary. Unfortunately, we can do nothing about the weather.

WING DESIGN

While the design of an efficient wing structure for private aircraft was briefly covered during the cost vs. design review (see Fig. 12-2 and 12-3), there are a few reasons why wings are designed as they are that will be of general interest. For example, we all know that tapered wings are more difficult to design and build since, among other things, all ribs are of different size. But when compared with a rectangular wing of similar span and area, tapered wings offer the dual advantages of (1) moving the center of lift inboard and at the same time (2) increasing the root chord. Moving the center of lift inboard reduces the root-bending attachment load because the moment arm is shortened; this in turn decreases the load that must be taken by the fuselage attachment and carry-through structure and so reduces structural weight. The increased root chord makes the beam deeper at the root, which further reduces attachment loads and structural weight. As another gain, the increased root chord provides more volume for main gear retraction. In addition to their favorable effect upon rolling rate as explained in Chap. 5, these are more reasons why tapered wings are used on high-performance aircraft. As an esthetic note, they also make a prettier airplane.

The safety advantages of locating all fuel out in the wings has been covered in Chap. 11; the structural advantages are presented in Fig. 12-7. This diagram also shows why root-bending loads decrease with a tapered wing as compared with a rectangular planform; since the center of the tapered area will be nearer the side of the fuselage, the moment arm will be shorter and the bending moment correspondingly lower as noted above.

Note that when the lift force or wing loading is up, the fuel and wing weights

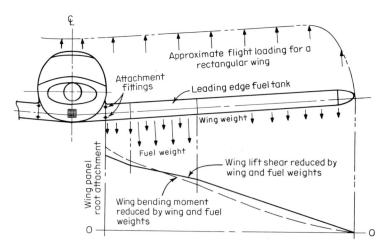

FIGURE 12-7
The effect of wing structure and fuel weights in reducing wing panel root attachment loads (rectangular wing shown).

act down to provide a reduction in root bending and vertical shear in accordance with

$$\text{Moment at wing root attachment} = M_w = (\text{lift} \times \text{lift arm})$$
$$- (\text{wing weight} \times \text{wing weight arm}) - (\text{fuel weight} \times \text{fuel weight arm})$$

$$\text{Vertical shear} = V_w = \text{lift} - \text{wing weight} - \text{fuel weight}$$

If the wing has more than one fuel tank per panel, the inboard tank or tanks should be emptied first if possible; working from the root out in succession provides the maximum reduction in wing bending and shear loads from the remaining outboard fuel. And this load relief acts throughout the flight envelope. When the wing receives a 3 g lift loading from gust or maneuver action, the load-relieving wing structure and fuel weights also increase by the same 3 g inertia factor (acting down to reduce wing bending and shear in the same ratio experienced in steady 1 g flight). Although the wing weight has been considered above as a concentrated load for simplicity, in actual analysis this weight would be distributed along the span in proportion to wing chord and skin thickness variation. Of course, landing gear and external stores fastened to the wing will also act to relieve wing bending and vertical shear loads.

Since the way wing loads are developed from airfoil coefficients has been covered in Chap. 2 and 6, with the effect of aspect ratio and flaps upon wing lift and drag also reviewed in Chap. 6, this information will not be repeated. But as a final wing-design consideration, every modern small aircraft should include flaps to improve short-field takeoff and landing performance without unduly penalizing cruise. For simplicity and effectiveness, the single-slotted flap provides excellent lift at moderate drag as shown by Fig. 6-4. Although the Fowler flap offers the greatest increase in lift, the mounting structure necessary to permit the aft and down extension movement required by this flap is quite complex in comparison with the fixed hinge point slotted flap. As a result, the true wing-area-increasing Fowler flap is seldom used on private aircraft.

TAIL, CONTROL SURFACE, AND FLAP DESIGN

Tail surfaces normally consist of four assemblies: fin, rudder, stabilizer, and elevator. On some aircraft the stabilizer and elevator are combined into one surface, referred to as an all-moving horizontal tail or stabilator.

In either case, the fin and stabilizer or stabilator usually have two beams: one located along the chordwise center of pressure of the surface area, with the other positioned just ahead of the hinge line.

Although there is nothing particularly exciting or novel about the design of tail and control surfaces, there are a few critical criteria. Most important of all being adequate *static balance* of the movable surfaces to prevent flutter, particularly for higher-performance aircraft. The amount of balance necessary may vary from none for low-speed aircraft up to 100 percent for high-speed designs. "No balancing" means the control surface has no weight added ahead of the hinge line to counterbalance the structural weight aft of the hinge line, while 100 percent balance means a surface is completely statically balanced about its hinge line. There is also need for *dynamic balance* of the movable surfaces, but this can only be determined by the vibration and flight tests briefly discussed in the following section on flutter.

From a design viewpoint, the ideal way to balance control surfaces is by incorporating a heavy box structure nose section ahead of the hinge axis with very light construction aft of the hinge line. Earlier aircraft accomplished this by use of fabric-covered light alloy frames, but to reduce maintenance this practice has given way to all-metal assemblies. Of course control surfaces also include ailerons, which can be quite critical in developing wing flutter unless adequately balanced and carefully flight-tested through V_D speeds.

Aerodynamic balance of control surfaces also becomes quite important at higher speeds or for relatively large control areas. Aerodynamic balancing is accomplished by locating part of the movable area ahead of the hinge line to reduce the surface loads that must be overcome by pilot control forces. Examples of this type of balance are shown in Fig. 7-3, where the area ahead of the hinge line acts to reduce surface forces aft of the hinge line.

Control surfaces provide greater lift per degree of displacement when the gap between the wing or tail structure and the control surface is sealed to prevent airflow between the high- and low-pressure sides of the control surface. Greater lift means a more responsive control system; but since gap sealing involves more assembly parts and work as well as some additional maintenance, sealed surfaces are usually found only on higher-performance and aerobatic aircraft.

Excellent background material for the design of aerodynamically balanced and sealed control surfaces is given in Refs. 12.5 and 12.6.

New aircraft designs might just as well incorporate cockpit-controlled trimming tabs on the elevator and rudder surfaces right from the first preliminary studies, since current FAA design specifications require flight demonstration proof that an airplane can be flown by use of these tabs if any part of the primary control system should fail (Ref. 12.7). While it seems difficult to imagine an in-flight accident that would sever any part of the primary control system without also damaging trim controls, it is possible to use friction-restrained vertical- and horizontal-moving trim levers to provide emergency elevator and rudder control response rates approaching primary control system levels.

Typical methods of light aircraft control surface and flap construction are shown in Fig. 12-8. Some modern fabrication techniques use a beaded surface rather than placing supporting ribs between load points, while some surfaces are bonded together under pressure instead of being riveted. Spotwelding is not commonly used for control surface assembly since this process is subject to fatigue and local corrosion in aluminum alloy structures, as well as being relatively expensive. But regardless of the construction selected, tail surface thickness should not exceed 12 percent for the reasons noted on page 94

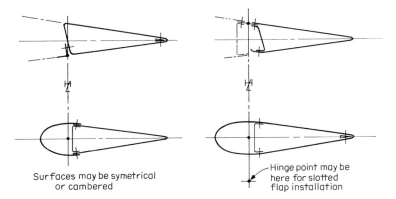

FIGURE 12-8
Typical structural design
for simple control surface and
flap assemblies

Surfaces may be symetrical
or cambered

Hinge point may be
here for slotted
flap installation

of Chap. 7, while the aileron and flap sections will be faired into the wing structure.

Those flat sides you may have noticed on control surfaces tend to keep the assembly from oscillating in flight by preventing flow changes due to curvature. To prevent surface flutter and nose-down trim from being induced by a drooping control surface, the simple piano-hinged sections shown in Fig. 12-8 should be at least 80 percent balanced about the hinge line when used as an elevator. This can be accomplished by a spring or concentrated weight mounted ahead of the hinge line (known as a bobweight), but must be flight-demonstrated to prove the system free from flutter for either design. In addition, if a spring balance is used the system must be free from flutter at V_D with the spring disconnected. So control surface design can be a bit more complex than it might appear at first. The proof of an acceptable system lies in flutter-free, responsive surfaces providing balanced control forces.

FLUTTER DESIGN AND TESTS

Since the subject of flutter has filled many books and can easily lead to much computer work, we shall merely review the causes and possible cures for this undesirable phenomenon which can occur in wing, tail surface, and fuselage structures.

What is generally referred to as "flutter" may actually be one of three somewhat related but different structural effects: actual flutter, divergence, or control surface reversal. To clarify these somewhat by definition:

1. *Flutter* is an unstable, self-excited oscillation of definite period set up in an airfoil and its associated structure by aerodynamic, elastic, and inertia effects which have combined to obtain energy from the airstream. Flutter will suddenly begin when the critical speed for a structure has been reached, and will increase in amplitude and intensity with increasing speed. So unless flight speed is immediately reduced, a fluttering structure may disintegrate. High-frequency flutter oscillations induced in heavy aircraft at high

speed will usually destroy the structure before any corrective action can be taken. Flutter is also more likely to occur at altitude where reduced density atmosphere exerts less damping effect upon the wing and tail structures. Tail flutter may also induce fuselage flutter in torsion and bending.

2. *Divergence* is the instability of a wing or tail surface in torsion, and may be critical when the center of lift is ahead of the center of twist (or torsional center) of a structure. For that condition, wing lift tends to increase wing angle of attack which further increases wing lift, etc. With a torsionally stiff wing, such as the D spar leading edge torque box design of Fig. 12-2, at speeds below some critical "structural divergence" speed the additional increments of twist with airspeed become smaller and equilibrium is obtained without any undesirable effects. In this condition the structure is said to be convergent.

3. *Control surface reversal* occurs when the fixed supporting surface is deflected in torsion (twisted) by the control surface in such manner that control response is opposite to that desired. For example, deflecting an aileron down produces a lift force on the wing structure which will normally be aft of the wing center of twist (or torsional center). If a wing is torsionally weak it will be twisted leading edge down, depressing instead of raising the wing panel.

   In similar fashion, a torsionally weak stabilizer may be deflected leading edge down when the elevator is moved down; this could provide quite a bit of activity in the cockpit since the airplane would abruptly nose up instead of trimming nose down as expected.

Rigid structures with minimum play in all hinges, control system components, and trim tabs are essential for a flutter-free airframe design.

Because all flutter-related structural deflections quite obviously result in very undesirable and dangerous flight handling characteristics, extensive ground, analytical, and flight tests are required to demonstrate that an airframe is free of flutter deflections at speeds 10 percent or more beyond the posted red line limit V_{NE} (never exceed speed = $0.9 V_D$ per Chap. 2 load review).

Ground vibration surveys are first conducted to determine whether any parts of the structure might be excited by engine operating frequencies and so start to flutter in flight. In the process we also have the opportunity to reduce airframe noise entering the cabin area. If you should ever see an airplane undergoing a vibration survey you will realize the level of airframe noise possible from engine excitation. These tests are usually conducted with low-power electric motors turning remotely driven eccentric weights, and there is no doubt when a section of wing, tail, and fuselage skin or structure becomes critical and "takes off." The resulting resonant vibration develops considerable noise from low power input, so just imagine what noise levels and structural deflections can be developed from 150- to 300-hp engines if they are not properly vibration-isolated or if the airframe structure is not sufficiently damped. To satisfy the important flutter-prevention requirements, all areas of critical vibration must be reinforced or damped in some manner to move their natural vibration frequency outside the excitation range. This is particularly important to assure dynamic balance of control surfaces prior to flight test.

Wing torsional deflection tests are also required by FAR Part 23 design specifications of Ref. 12.8 and must satisfy certain empirical criteria intended to preclude divergence and control surface reversal. For those readers interested in flutter, Ref. 12.9 provides excellent comprehensive design sources covering flutter prevention as well as approved ground and flight test procedures.

Since flutter can be critically affected by control surface balance, there are a few maintenance items that should be observed as part of every pilot's anti-flutter campaign. Mud, snow, slush, or ice accumulation on control surfaces will upset their balance and should be completely removed during preflight inspection. In similar manner ice, dirt, and dust can become lodged inside a surface aft of the hinge line and destroy the existing balance; dirt and dust buildup should be checked during every 25-hr airframe inspection, with the possibility of interior ice checked during every cold-weather preflight.

If flutter should occur from any of the above deposits or a local structural failure, reduce power at once and bleed off airspeed to some level about 30 percent above stalling speed. Be extremely careful to retain a comfortable margin of control speed plus coordinated controls with the ball centered, and land as soon as possible. Do not ever attempt to continue flight with a structure ready to flutter if given the opportunity—which will surely occur as soon as critical flight speed has been reached, and then probably for the last time.

In addition to clarifying just what causes flutter and indicating methods for prevention, this brief review has been included to let you know that every certified airplane has been well tested to prevent any possibility of flutter when operated within the approved weight, balance, and flight envelopes as originally designed and certified.

COMPOSITE STRUCTURES

Purchase price has such an important effect upon commercial sales that increasing amounts of design effort are directed toward development of composite structural materials to reduce airframe cost. Combinations include fiber glass or other synthetic fabrics bonded to various rigid foam materials; graphite, boron, or similar filament fibers in epoxy-type binders; honeycomb core materials with metal or impregnated synthetic fabric facing; chopped fibers, cloth and synthetic resins; etc. Reference 12.10 provides an excellent review of these possible combinations, including production methods and their effect upon reducing aircraft cost through automation.

Interestingly enough, the new Bellanca Aircraft Engineering Skyrocket II of Fig. 12-9 has an airframe fabricated from epoxy-prepreg composite material and aluminum honeycomb. In addition to providing a structure that can be assembled by relatively unskilled labor, the resulting synthetic surface is aerody-

FIGURE 12-9
Bellanca Aircraft Engineering
Skyrocket II
Bellanca Aircraft Engineering Photo

namically superior to that of aluminum alloy skin and rivet construction—hence the world speed records won by Skyrocket II.

Further work will be necessary to obtain FAA acceptance of composite structures, however, since strength repeatability can vary from airframe to airframe depending upon the assembler, the amount of bonding resin used, humidity at time of assembly, dirt in the air, etc. As a result, strength margins must be increased for composite structures, adding to the empty weight and materials cost. Startup costs will also be greater since rather expensive tooling and fairly sophisticated quality assurance procedures are necessary before the advantages of employing relatively unskilled labor can be realized.

If we assume all this can be accomplished, the crash and lightning protection provided by a composite structure is considerably less than that of a comparable sheet metal assembly. Unless new composite techniques can be demonstrated to offer the same level of passenger safety available with sheet metal construction, it is doubtful that the flying public will accept composite structures—plastic aircraft—regardless of any accompanying cost or performance advantages.

REDUCING CABIN NOISE

Since the powerplant is the primary source of cabin noise, we need improved methods of engine suspension and isolation that will totally prevent engine vibration from getting into the airframe. Once lightweight skin panels become excited by engine frequency, considerable noise is developed as noted in the flutter discussion. Heavier skins or dampened skin panels will lower the cabin noise level, while dynafocal engine mounts are quite effective in isolating engine vibration during flight.

The need for a well-damped structure also applies to window areas, which require thick acrylic or similar plastic to provide an effective noise barrier. The one remaining weapon against interior noise is the most obvious—soundproofing. And the frequently overlooked firewall area should be a prime candidate for considerable soundproofing protection. If properly applied, modern soundproofing materials will also provide insulation and panel-deadening (vibration-damping) properties which are most desirable.

Unfortunately, but typical of aircraft development, these improvements are accompanied by increased weight and cost. In return, the resulting airframe should be quieter and the airplane considerably safer for long trips due to reduced fatigue and improved cabin communication. And what more can we expect than a safe structure and quiet airplane—and we should not have to accept less at the present state-of-the-art of aircraft design.

REFERENCES

12.1 FAR 23.572 (a) (2).

12.2 FAA Advisory Circular 43.13-2, pp. 45–56.

12.3 Karl Bergey, *The Design of Crashworthy General Aviation Aircraft.* Society of Automotive Engineers paper 740376, April 1974.

12.4 FAR 23.349 (a) (2).

12.5 L. E. Root, "The Effective Use of Aerodynamic Balance on Control Surfaces," *Journal of the Aeronautical Sciences,* April 1945, pp. 149–163.

12.6 Hoerner, "Aerodynamic Drag." Published 1951.

12.7 FAR 23.677 (b).

12.8 FAR 23.629.

12.9 (a) J. E. Younger, *Structural Design of Metal Airplanes,* chap. XV, pp. 271–277. McGraw-Hill Book Company, New York, 1935 and subs.

    (b) W. S. Diehl, *Engineering Aerodynamics,* chap. 7, pp. 228–234. The Ronald Press Company, New York, 1939 and subs.

    (c) ANC-12, *Vibration and Flutter Prevention Handbook,* April 1948 as revised.

    (d) *Simplified Flutter Prevention Criteria for Personal Type Aircraft.* Report No. 45, Airframe and Equipment Engineering Branch, Department of Commerce (now the FAA), as corrected to February 1952.

    (e) Flight Test Report Guide, pp. 70–73; FAA, May 1969.

12.10 (a) San Diego Aircraft Engineering, Inc., *Potential Structural Materials and Design Concepts for Light Aircraft.* NASA CR-1285, March 1969.

    (b) A. C. Marshall and J. Brentjes, *Use of Honeycomb and Bonded Structures in Light Aircraft.* Society of Automotove Engineers paper 730307, April 1973.

    (c) H. D. Neubert and R. W. Kiger, "Modern Composite Aircraft Technology," *Sport Aviation,* July, September, and December 1976.

13 EQUIPMENT REQUIREMENTS

Although the preceding chapters have been primarily concerned with handling, performance, and structural aspects of aircraft design, these are all rather nebulous or hidden features inherent in the basic airplane. What seriously affects pilots' comfort and safety as well as their opinions of a new model are those items used on every flight—the equipment.

Equipment includes seating and harness, all controls, instruments, radios, and such major systems as airspeed, electrical, hydraulic, vacuum, oxygen, and, by no means least, cabin heating, defrosting, and ventilating (which may include air conditioning). As long as the seating and control locations are comfortable and all systems are functioning well, we have a "good airplane" and happy crew. But let a radio operate intermittently, the battery spill over, or one omni go out during an IFR hold and the entire flight goes sour. Now the airframe, engine installation, and handling characteristics have not changed one bit, only one or possibly two pilot contact items have failed; but since this experience directly affects the operator, it will be remembered as an extreme annoyance—amplified by thoughts of the installation cost.

Aircraft designers all too frequently concentrate primarily upon airframe and powerplant details, throwing in equipment at the last minute to balance the total package. Poor ventilation is a classic indication of this approach although recent designs have shown some improvement, particularly for windshield defrosting. While it is obviously important to have a sound airframe and powerplant design, all this is of no use if the resulting airplane is inconvenient and uncomfortable.

With some hope of motivating future improvement, this chapter offers suggestions permitting owners to correct equipment areas and items that may be found unsatisfactory, while also covering design features essential for reducing maintenance. In this regard, accessibility and reliability are important considerations for every equipment installation. In practice this means if maintenance time and charges are to be kept low, equipment must be accessible; if an item of equipment will be used and depended upon for flight safety, it must be accurate and reliable. With these objectives in mind, let us briefly consider principal equipment features and operating systems.

CABIN DESIGN All equipment relates in some way to pilot and passenger accommodation. Since our coverage must be rather broad and general in nature, a review of operational priorities will assist in highlighting major differences in basic equipment requirements.

Aircraft intended for the serious business and pleasure markets must have as much elbow room as possible for all occupants along with well-balanced and friction-free controls, a heated Pitot system, and ample panel space for the instruments and avionics necessary for IFR operation. Adequate ventilation, cabin heat, windshield defrosting, seat belts, and shoulder harnesses are also essential for passenger comfort and safe operation.

Sport and aerobatic aircraft have quite different requirements in that a tight cockpit plus strong restraint system are necessary for pilot safety and to provide the level of handling ability realized by "strapping on the airplane." And well-balanced, low-friction controls are equally important for proper "feel" and control response. If the design is an open cockpit type, nature provides the ventilation. Since IFR operation would not normally be expected of aerobatic aircraft, the instrument panel and avionics could be minimum—although an omni would seem useful for cross-country flying or an occasional VOR (very-high-frequency omnirange) approach in unexpected weather.

Agricultural aircraft are pieces of flying farm machinery, but this does not mean Spartan accommodation should be the design standard. Ag pilots may fly in extreme heat for well over 10 hr a day during seasonal peaks. As a result, elimination of pilot fatigue becomes a critical consideration, with balanced and highly responsive controls being essential to reduced pilot workload. Modern Ag planes will include cockpit air conditioning to eliminate heat fatigue and to more fully exclude spray and dust chemicals, thereby literally reducing cockpit irritation. Since being cool is not the only criterion for comfortable long duration operation, increased attention is being paid to seat design for both pilot comfort and crash protection. And of course, rugged seat belt and harness restraint systems are especially important for this type of activity. Again, complex avionics and instrument systems are not important, although reliable and accurate stall warning, airspeed, altitude, manifold pressure, rpm, and fuel level indicators are necessary for safe operation. Since Ag planes infrequently fly cross-country, many are equipped with only a portable communications radio.

To round out the picture with another major design variation, sailplanes have special requirements unique to their type. Powerplant instruments are replaced by variometers to indicate rates of climb or sink in thermals, and a snug cockpit with heavy-duty shoulder harness and seat belt is mandatory for higher-performance models to prevent pilot contact with the canopy in rough air as well as to provide off-field landing protection. To reduce frontal area drag, the reclining cockpit design is now standard for all higher-performance sailplanes. Complex avionics are not required, although a simple omni would seem advisable for distance flying and orientation on hazy days. Since sailplanes are not permitted IFR flight during contests per Soaring Society of America rules, there is no need to include panel space for various gyros and navigating equipment items; but an electric turn and bank could be helpful in hazy day flying. The surface controls of higher-performance sailplanes in particular must be free of any friction since control forces will normally be extremely light, reducing "pilot feel" to minimum levels during most flight conditions.

From this review it is apparent that a good flight control system is basic to all aircraft. Such a system is one in which (1) all control forces are evenly balanced and feel similar for elevator, aileron, and rudder operation; (2) all controls are sufficiently free of friction to permit the surfaces to return to neutral when released in flight; and (3) the surfaces provide positive and rapid control response about all three axes.

Balanced control forces will depend upon required control surface areas, the degree or extent of aerodynamic balance used, and the actual geometry of the individual control systems. By working back through each control system using the applied loads calculated for each set of control surfaces, it is possible to determine the maximum control forces that would be expected in flight at maximum speed. Naturally all control forces will be lower at lower speeds, but they will reduce in proportion since the same airflow passes over all surfaces. By this procedure it is possible to design for balanced control forces, although some adjustments will undoubtedly be desirable following initial flight tests.

Ball- and needle-bearing pivots offer low-friction control systems, while the criteria for responsive control surfaces have been well covered in Chaps. 5 and 7.

All engine and flight controls must be located convenient to the pilot. Flight controls should be designed to clear legs and lap to permit full control motion without restriction, and to provide a clear spot for charts or a flight board. Even though many of us believe a stick offers more positive control, the wheel has gained popularity primarily because it does clean up the cabin; although it is a heavier and more expensive installation than a control stick.

You may be reassured to learn that although actual control surface design loads may be comparatively low, modern FAA practice requires the primary control systems—which are the elevator, aileron, and rudder controls—to be designed for maximum pilot effort. Known as a "jamming load condition" on the premise that a pilot could exert such forces if a surface became locked or jammed in a fixed position, the pilot limit load forces are:

a. Aileron control—stick 67 lb; wheel 50 lb, applied at the rim of the wheel on each side and in opposite directions

b. Elevator control—stick 167 lb; wheel 200 lb

c. Rudder control—200 lb per pedal

These forces must be applied left and right, push and pull upon stick or wheel based upon load conditions set forth in Ref. 13.1. All values must be multiplied by a factor of 1.50 to obtain ultimate loads for detail design. Since control system strength must be demonstrated by static load test for each newly certified design (Ref. 13.2), control systems rarely cause any problems other than those of routine service maintenance. To simplify the required 100-hr and annual inspections, access handholes or hinged panels should be provided at all major control system pivot, pulley, and adjustment points.

Because visibility is something impossible to have in excess, an unlimited number of different arrangements have been tried in attempts to improve visibility from the pilot's seat. Working upward from the periscope used on Lindbergh's *Spirit of St. Louis* of the late twenties and Hawk's retractable cockpit-enclosure-equipped *Time Flies* of the late thirties, we find the ultimate in visibility on some current homebuilt designs such as the *Breezy* where the pilot sits right out in the open air in best Glenn Curtiss tradition. Somewhere in between lies the desirable balance of at least 240° visibility around the horizon (in azimuth) and 11 to 13° down over the nose. The former is important for visually sweeping the area before starting a turn to either side, while the downward visibility is important during taxi, climb, and approach phases of flight.

The 11° downward vision angle is also quite useful in another way: from an altitude of 3000 ft above ground level, this angle permits sighting the ground almost exactly 3 mi ahead of the airplane. So when the ground disappears in haze, fog, or rain, turn around—you have automatically been advised that conditions ahead are below VFR (Visual Flight Rules) minimums, and with sufficient warning to permit a safe 180° turn while still VFR. The downward and forward vision angle may be readily checked from your normal seat position by leveling the plane to flight attitude, sighting the nearest ground point, and then computing the downward angle based upon eye height above ground per Fig. 13-1. Any time taken to determine this in-flight visibility indicator will be well spent for additional flight safety.

Since cabin visibility means safety in many ways, this becomes an important design consideration. Cessna has provided exceptional flight visibility for a

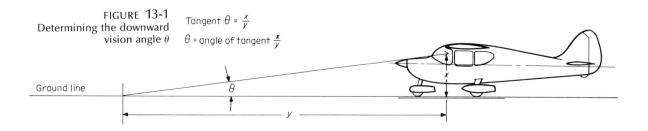

FIGURE **13-1**
Determining the downward
vision angle θ

$\text{Tangent } \theta = \frac{x}{y}$

$\theta = \text{angle of tangent } \frac{x}{y}$

Ground line

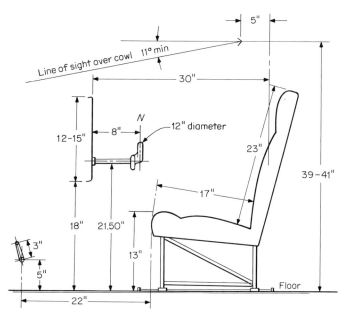

FIGURE 13-2
A typical cabin arrangement
of seating and controls

series of single-engine aircraft in which the upper cabin offers an almost 360° sweep as shown by Fig. 1-3, and the vertically adjustable pilot's seat permits satisfactory downward vision over an otherwise high instrument panel.

Pilot seats mounted on tracks that incline up as the seat comes forward should be lighter and much cheaper than the present vertically adjustable ones, and so deserve further consideration. As another design approach, military seats move vertically on guides (as necessary for ejection) while the pedals adjust fore and aft to accommodate different leg positions. Having both pedals and seats adjustable would be very nice for changing position on long flights and probably is the ultimate arrangement. But because these features increase cost and weight, only movable seats are normally used in small airplanes; although adjustable seats and pedals are quite commonly included on high-performance sailplanes.

Figure 13-2 shows a typical, comfortable modern cabin and control system arrangement. Any additional rows of seats should be separated by at least 32 in. from those ahead and preferably by 36 in. For readers interested in developing cabin and cockpit accommodations, the 1/16 scale person and seat arrangement included as Fig. 13-3 will be found useful for preliminary layout and study.

All seats must be designed to carry at least a 970-lb vertical load, with aerobatic seats stressed for 1710 lb of passenger and parachute per Ref. 13.3. In addition, each seat and supporting structure must be tested to withstand crash loadings of 9 g forward and 1.5 g to either side.

Seat belt load and installation requirements are similar to those specified for

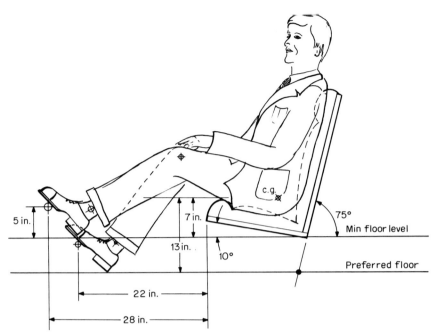

FIGURE 13-3
A ¹/₁₆ scale man, seat, floor,
and pedals diagram

seat design. Seat belts and harnesses must be tested to FAA Technical Standard Order (TSO-C22) and approved by the FAA for aircraft use. Over the years, a great number of materials and release system improvements have gone into seat belt development; if you have ever spun or been upside down in an open cockpit airplane, you appreciate just how comforting and important a strong lap belt can be.

Today's FAA design requirements place equal emphasis upon seat belt and shoulder harness installations for crash protection. World War II clearly showed a reduction of about 50 percent in crash fatalities when harnesses were used, so always use yours. While no one expects to crash, there is precious little time to strap on a harness when weather or other problems dictate a rough field landing. For readers interested in harness requirements, attachment loads were noted in the fuselage design section of Chap. 12, with installation details covered by Ref. 12.2.

Thanks to the design requirements of current FAA specifications, when well-strapped and harnessed in place you are able to withstand anything the fuselage can—and many tests will have been performed to prove this is so. As becomes increasingly evident, with all the test work required for certification it is no wonder development costs climb faster than the airplane ever will.

Instruments, instrument location, and lighting have probably varied over the years more than any other equipment item. Of course, much of this change resulted from technical advances in instrument quality and reliability. But an equal amount of variation in type and location occurred through individual preference, every pilot having his or her own idea of the best possible panel layout.

As aircraft came into "mass" production and rental ownership developed, flight safety was frequently related to the speed with which pilots could locate specific panel instruments in an emergency. Following a brief pattern checkout they might be flying solo in an airplane new to them, with instruments not at all similarly located to those of the plane flown for the past 30 or 40 hr. And so corrective action could be slow when needed and frequently too late.

To improve this situation a standard location of the basic IFR flight instruments is now part of the certification requirement for aircraft over 6000 lb gross. This same arrangement is applied by the FAA to all aircraft equipped with attitude and directional gyros, resulting in the familiar T panel layout of Fig. 13-4.

To assist in reducing familiarization time between different airplanes, whether homebuilt or production types, this instrument arrangement should be followed for all models equipped with gyro instruments. Note that the attitude and directional gyros are specified as vacuum or pressure driven, while the turn and bank (T & B) is electric. This practice eliminates a second air-regulating valve for the T & B while also providing dual lateral attitude indication—a backup system—to prevent those critical spiral dives should an electrical or vacuum system fail. If the vacuum system goes out, it is possible for IFR-trained pilots to maintain a level and safe attitude by scanning the (electric) T & B, airspeed, and ROC instruments. In fact, pilots should frequently check their proficiency in maintaining level flight with these three instruments, but under flight instructor supervision. With practice it is possible to maintain level flight and altitude with just the T & B plus airspeed operating, so the basic IFR instrument cluster of Fig. 13-4 provides overlapping information when all systems are operating.

The most important single instrument system is the Pitot-static (or airspeed)

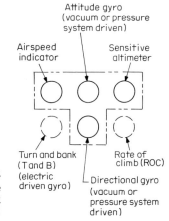

FIGURE 13-4
Instruments arranged in the T panel layout of FAR 23.1321

installation. This system is basic to aircraft operation because it provides information for the altimeter and ROC instruments in addition to indicating flight speed. For FAA certification, airspeed indication must be accurate to within 3 percent or 5 knots, whichever is greater, of true airspeed. Airspeed calibration accuracy is thoroughly checked as part of the certification flight test program, with corresponding airspeed conformity checks performed as part of the flight test procedure for every production airplane. So if your Pitot-static system is kept free of bugs, mud, ice, and water, it will indicate flight speed quite accurately.

We have four ways to ensure reasonably accurate Pitot-static system operation in addition to the normal preflight inspection, with this number of options alone being an indication of the importance assigned to the primary airspeed system. These are (1) water drain traps which are bled to remove any condensation or rain that has entered the pressure lines and could result in false readings, (2) an electrically heated Pitot-static (airspeed) head to prevent ice accumulation, (3) an alternate static source that may be opened if the primary static pickup is blocked, and (4) a variation of the alternate static which is to break carefully the ROC face to permit emergency static pressure to enter the system for altitude indication. ROC readings must then be disregarded since this instrument will be working in reverse; use the altimeter as a climb reference if necessary.

The Pitot and static line water drains should be accessible in flight, if possible, just in case trouble is experienced during flight through a severe rain shower. These quick drains must be located at the lowest point in the airspeed lines to permit easy inspection at least every 25 flight hr, or during preflight whenever the airplane has been tied down in blowing rain.

An electrically heated Pitot-static head is highly recommended for all aircraft since it not only permits removal of unexpected ice accumulation from the airspeed pickup, but heat may be helpful in removing water as well. If only one option is possible, the electrically heated airspeed head is preferable to an alternate static inlet because there is no way to open an iced-over Pitot head without heat, while altitude indication may be obtained by breaking into the ROC case as previously mentioned. Or stated another way: When both the Pitot and static pickups are thoroughly out of action due to ice accumulation, the alternate static will be of no use for airspeed indication because there is no dynamic (Pitot) pressure available. Of course, an alternate static will provide acceptably accurate altitude and ROC indication, so we find many modern aircraft equipped with both a heated airspeed head and an alternate static source.

During night operation lighted instruments must not create confusing reflections all over the windshield. Individual post lights mounted through one of the instrument attachment screw holes provide excellent lighting when all panel lights are wired into a single dimming switch. Any instrument reflections can be contained by mounting a padded shield off the instrument panel top deck, being careful not to block instruments or switches from pilot view.

As a final instrument note, an audible stall warning indicator is practically mandatory for all newly certified aircraft. Safe Flight Instrument Corporation is the leading supplier of a small wing-mounted pickup used on all modern United States airplanes. This installation satisfies FAR Part 23.207 requirements by providing both a red warning light and buzzing signals throughout an adjustable speed range prior to full stall.

If you have instrument work to be done, necessary design and operational requirements for acceptable airspeed indicating systems are given in Ref. 13.4, while practical instrument installation and maintenance suggestions have been gathered in Ref. 13.5—many of which can be readily performed by private owners.

While opening a door or sliding back the canopy on closed cabin aircraft are about the only steps the average owner can take to reduce stifling ground heat on a sticky midsummer day, air-conditioning options are now available for some modern aircraft. Whether or not this solution with its attendant weight and cost is worthwhile will depend upon personal preference and home base location. Certainly our northern areas do not warrant air conditioning, but this equipment must be a great source of relief during southern summers.

Whether air conditioned or not, closed cabin areas must be provided with reliable sources of cold and hot air during flight. Air delivery should be sufficiently quiet not to interfere with radio communications or become a source of fatigue, with outlet flow broadly disbursed and not ejected as if from a fire hose on whoever is nearest each opening. Heat is normally added to fresh outside air by passing through a transfer muff carried as part of the engine exhaust system (Fig. 11-3), and may be selectively mixed with cold air to provide the desired degree of cabin heating.

One ventilation feature often neglected is the frequent need for windshield interior defrosting prior to takeoff. Since there is no moving air in most ventilating systems until airborne, there can be no defrosting without an auxiliary blower—and this is a badly needed item of equipment. Because windshield condensation is quite common during summer and fall operation in mountainous terrain and on the water, and a regular pre-takeoff experience in wintertime, provision should be made for pilot selection of heated defrosting air during ground operation. To be acceptably adequate, ventilation systems should include a windshield defrosting blower as standard equipment.

ELECTRICAL SYSTEMS   Aircraft electrical systems are usually taken for granted by owners and pilots, in much the same manner as for an automobile; get in, turn on the key switch, and start off. This may be sufficient for VFR day sport flying where the most important job an electrical system performs is reliably starting the engine. And even that assistance is not expected on some of the older hand-cranked classic models.

For cross-country flying where reliable communications, navigation, and various forms of lighting are necessary, a dependable electrical system is essential. For our purposes, this means the basic wiring and any subsequent additions must be performed in accordance with acceptable standards of wire size for the load carried and length of run; with proper switch, fuse, and circuit breaker size; and with reliable electrical accessories, such as rheostats, Pitot heat, navigation lights, panel lights, indicator lights, rotating beacon or strobe lights, and landing lights.

Adding in the radios and related navigating equipment results in a total load of many amps when all items are operating during night IFR conditions. To make sure everything keeps running at maximum demand, FAA certification requires an electrical load analysis proving sufficient alternator or generator capacity is available to satisfy peak amperage. Slight overload can be supplied from the battery for short periods of time—for example, turning on landing lights during final—but load requirements should preferably not exceed 80 percent of alternator or generator capacity at the maximum continuous running load.

A typical electrical load analysis is simply compiled from equipment item specifications and would follow this format (for a 12-v system):

Item	Maximum amperage
Fuel quantity indicators	0.90
Oil temperature gauge	0.30
Oil pressure gauge	0.30
Fuel pressure gauge	0.30
Ammeter	0.30
Auxiliary electric fuel pump	0.70
Landing gear position indicators (2 lamps)	0.24
Heated Pitot-static head	12.00
Dome lights (**)	1.00
Panel lighting—lamps and rheostat	1.04
Internal lighting in compass, T & B, radios	3.00
Navigation lights—wing tips and tail	4.04
Strobe lights (2)	2.40
Landing lights (**)	6.00
T & B (electric)	0.75

Stall warning system	0.12
IFR radio equipment (dual nav., dual com., glide slope, ADF, transponder, DME, and audio panel)	13.00
Load with all items operating	46.39 amps
10% overload safety margin	4.64 amps
Total maximum condition	51.03 amps

The items marked (**) are not required for normal cruising flight (long period) operation and so may be discounted. We then have:

Load with (**) items turned off (46.39 − 7.00)	39.36 amps
10% overload safety margin	3.94 amps
Total amps for system design	43.30 amps

To supply this system demand at 80 percent output the alternator or generator must be rated at

$$\frac{43.30}{0.80} = 54 \text{ amps capacity}$$

so a 60-amp, 12-v alternator or generator would be adequate. Similar load analysis will determine the electrical source capacity required for new or modified systems.

As military aircraft grew larger and more complex, 24-v systems became standard, principally to reduce power losses throughout the system. Most small aircraft have standardized on 12 v, now common on automobiles as well, and so recommended wire size, circuit breaker or fuse ratings, and switch charts have been included as Fig. 13-5 to cover 12-v electrical systems. The wire run vs. size data may also be safely applied to 24-v systems at half the indicated amperages, since the line losses would be lower than for 12-v runs of comparable length. These data should enable any owner to plan wiring changes that will not overheat, although the system load must be rechecked if additional electrical equipment is to be installed and operated during flight. Any owner feeling up to the task can make many types of wiring changes under the supervision of approved mechanics—and at a considerable saving over repair shop costs.

Normal aircraft wire should be of MIL-W-16878 or MIL-W-5086 quality, with the former being superior for higher temperatures and salt-water exposure. Magneto wire should be shielded to meet specification MIL-W-76B or better, with generator and mag filters installed to reduce radio interference noise.

While battery location may really depend upon balance requirements, since

Wire size	20	18	16	14	12	10	8	6	4
Stranding	7 × 28	16 × 30	19 × 29	19 × 27	19 × 25	19 × 23	19 × 21	49 × 23	49 × 21
Circular mil area	1094	1568	2340	3777	5947	9443	15105	24354	38430
Circuit current, amps	MAXIMUM LENGTH OF CONDUCTOR IN FEET FROM POWER SOURCE TO LOAD AND RETURN								
1.	36.4	52.3	78.0						
2.	18.2	26.1	39.0	63.0	99.0				
3.	12.2	17.4	26.0	42.0	66.0				
*4.	9.1	13.1	19.5	31.5	49.5	78.8			
5.	7.3	10.4	15.6	25.2	39.6	63.0			
*6.	6.1	8.7	13.0	21.0	33.0	52.5	83.8		
7.	5.2	7.4	11.1	18.0	28.2	45.0	72.0		
8.		6.5	9.8	15.8	24.8	39.4	63.0		
*9.		5.8	8.6	14.0	22.0	35.0	56.0	90.0	
10.		5.2	7.8	12.6	19.8	31.5	50.4	81.3	
*15.			5.2	8.4	13.2	21.0	33.6	54.1	85.5
*20.				6.3	9.9	15.8	25.1	40.6	63.4
*30.					6.6	10.5	16.8	27.1	42.7
40.						7.9	12.6	20.2	32.0
50.						6.3	10.1	16.2	25.6
55.							9.2	14.7	23.3
60.								13.5	21.4
75.								10.8	17.1
90.									14.2

(a) Stranded conductors for 12-volt circuits, 3 percent voltage drop.

Amperage	
Normal rating	Maximum continuous load
4	3.6
6	5.4
9	8.1
15	12.6
20	17.0
30	24.0

(b) Circuit breaker and fuse loadings.

FIGURE 13-5
Electrical system design data: wire length vs. load, circuit protection, and switch selection.
(a) and (b) from the Boating Industry Association; (c) and (d) from AC43.13-1A, p. 177.

every battery represents a fairly substantial mass of lead ballast, it is important that this installation be accessible for frequent inspection and service. Every battery, whether self-contained and vented or requiring a battery box, must be vented overboard at a location that will neither siphon the battery nor spill upon the exterior (and ruin the finish). As a structural protection, the area immediately around and below a battery should be coated with a bitumastic-type compound resistant to battery acid. Battery size will depend upon starter and engine requirements, but for colder climates a 37 amp-hr capacity should be the minimum for a 12-v system starting 150 to 200 hp.

Since Refs. 13.6 and 13.7 provide electrical system design requirements and useful maintenance suggestions, interested readers may consult these sources for more detailed electrical information.

RADIO SYSTEMS   The FAA design regulations say very little about radio installations—and really don't need to since pilot opinion develops quite strongly and rapidly whenever radio equipment operates intermittently or fails completely. There is nothing more disturbing than an ADF (automatic direction finder) that won't pick up a

(c) Wire and circuit protector chart.

Wire AN gauge copper	Circuit breaker amp	Fuse amp
22	5	5
20	7.5	5
18	10	10
16	15	10
14	20	15
12	25 (30)*	20
10	35 (40)*	30
8	50	50
6	80	70
4	100	70
2	125	100
1		150
0		150

* Figures in parentheses may be substituted where protectors of the indicated rating are not available.

(d) Switch derating factors.

Nominal system voltage	Type of load	Derating factor
24 V DC	Lamp	8
24 V DC	Inductive (relay-solenoid)	4
24 V DC	Resistive (heater)	2
24 V DC	Motor	3
12 V DC	Lamp	5
12 V DC	Inductive (relay-solenoid)	2
12 V DC	Resistive (heater)	1
12 V DC	Motor	2

NOTES:

1. To find the nominal rating of a switch to operate a given device, multiply the continuous load current required by the device by the derating factor corresponding to the voltage and type of load.

2. To find the continuous load current that a switch of a given nominal rating will handle efficiently, divide the switch nominal rating by the derating factor corresponding to the voltage and type of load.

FIGURE 13-6
Modern antenna installation
on the Rockwell Commander
Rockwell International-General
Aviation Division Photo

strong beacon 6 mi away today but unerringly homes in on a weak one tomorrow.

It is common practice to include all communicating, navigating, and antenna items under the general terminology of radio (or avionics) equipment. Modern solid-state radios are available at all levels of complexity and reliability, with prices increasing right along with the degree of sophistication. While model selection is a matter of personal opinion and operating requirements, it does not seem necessary to purchase TSOd equipment for normal use, including IFR operation if dual nav/com units are installed.

In view of the great amount of material available concerning radio and antenna installations, a few of the more useful having been included as Ref. 13.8, we shall cover only a few pertinent items in this chapter. It may help to know it is possible to drive two omnis plus a glide slope receiver from one V or "balanced loop" antenna mounted on the fin or fuselage. The antenna lead-in will go directly into a small "black box" known as a *triplexer,* which divides the signal received into components required by each omni plus the glide slope receiver.

If only two omnis are used, or one omni with glide slope, the single antenna lead-in would go into a *diplexer* which converts the incoming signal into two parts for omni and glide slope presentation. The point being that even when three items of navigation or instrument landing system equipment are used, all can be fed from one omni antenna by use of *couplers,* as the diplexer and triplexer are known. So two or three receiving antennas are not necessary, although they will be required if couplers are not used.

Whether installing new or add-on equipment, a small ball-ended transponder-type antenna may be used quite successfully as a DME antenna rather than the more familiar and expensive blade type. When dual communications sets are installed two com antennas are mandatory; for best reception these should be mounted more than 2 ft apart and not too close to the ADF sense antenna. Figure 13-6 shows a well-planned, modern antenna system suitable for small-aircraft IFR avionics.

Back in the cockpit, the DME package must not be panel-mounted too close to the ADF receiver because it may emit energy interfering with ADF reception and cause pointing errors. Otherwise the avionics equipment seems to be quite compatible, provided all items are properly wired.

When more than one nav/com system is installed along with an ADF and transponder, DME, or marker beacon, a single control panel should be used to bring everything together and also permit instant switching between preselected frequencies. This convenience is most useful when departing a busy airport, communicating en route, or making an IFR approach.

Communication and navigating equipment that operates unreliably and cannot be depended upon for accuracy is worse than no equipment at all. In view of this it is necessary that all radio items used for serious IFR operation be properly installed with adequate cooling, equipped with recommended antenna and lead-in accessories, and maintained in satisfactory condition.

VACUUM/PRESSURE SYSTEMS

Some type of power source is required to operate gyro-driven instruments. To provide instrument redundancy, current practice prefers the T & B indicator to be electrically driven, while the attitude and directional gyros are powered by an engine-mounted vacuum or pressure system pump.

The pressure system is preferred for aircraft operating at higher altitudes or with pressurized cabins, while most of our lower-performance general aviation models are presently powered by vacuum systems. Most new installations will probably be of the pressure type since this system provides greater mass airflow at altitude, resulting in better pump cooling, higher efficiency, and improved gryo life. It also offers an installation with less plumbing as shown by Fig. 13-7.

While the owner's manual or manufacturer's data should be checked for recommended operating limits, most gyros are designed for 4 to 5 in. of mercury

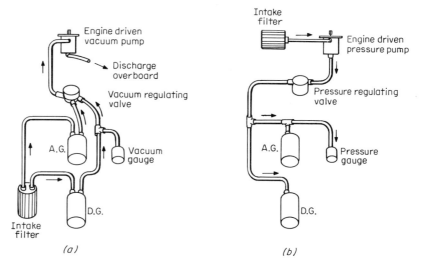

FIGURE 13-7
Vacuum and pressure systems suitable for gyro operation.
(a) Schematic layout for a typical vacuum system.
(b) Schematic layout for a typical pressure system.
Based on Airborne Manufacturing recommendations

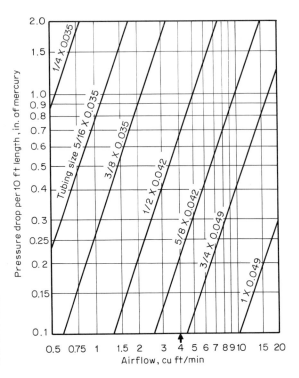

FIGURE 13-8
Line pressure drop chart for
smooth tubing used for
vacuum and pressure pump
systems
After AC43.13-2, Fig. 11.1

differential pressure or vacuum, with 4 in. considered about the optimum for attitude and directional gyros.

Anyone interested in installing or relocating gyro equipment will find basic design requirements in FAR 23.1331, including the need for two separate instrument power sources for multiengine aircraft. Reference 13.9 offers useful installation data and sample calculations for adding gyros, while Fig. 13-8 may be used as a plumbing guide. Incidentally, the average total air consumption of 4 cu ft/min for the attitude and directional gyros has been noted on this chart.

HYDRAULIC SYSTEMS

Although most fixed gear small aircraft do not have complex hydraulic pump, selector valve, and piston actuated hydraulic retraction systems, they do have one hydraulic system nevertheless—the brake system. And as required for all hydraulic installations, brake systems must be static tested at 1.5 times the maximum operating pressure per FAR 23.1435.

Since hydraulic system design, installation, and maintenance have been thoroughly reviewed in many standard reference books in addition to the coverage of Ref. 13.10, detailed discussion does not seem warranted here. But in the interest of reducing system weight, cost, and maintenance it is suggested that designers consider a double-acting hydraulic hand pump and accumulator for the source of hydraulic power. Pumping up the accumulator while en route would build up a reservoir of pressurized fluid sufficient to lower and positively lock the landing gear upon entering the pattern. Similarly, hydraulic pressure could be stored up before starting the engine, permitting instant gear retraction after takeoff when the selector valve is moved to "wheels up" position. After all, the operating cylinders respond to pressure built up by muscle power just as fast as they do to pump power, and the simpler system results in fewer hydraulic headaches plus lower maintenance costs.

OXYGEN REQUIREMENTS

Although few of us now fly at altitudes requiring oxygen assistance, such operations will increase with growing use of turbocharged powerplants. Present rules require the flight crew to use supplemental oxygen if remaining at cabin altitudes between 12,500 and 14,000 ft mean sea level (MSL) for periods exceeding 30 min. Operation at cabin levels above 14,000 ft MSL is prohibited unless at least the flight crew uses oxygen at all times; and every occupant must be provided with supplemental oxygen at all cabin pressure altitudes over 15,000 ft MSL (FAR 91.32).

Certification of supplemental oxygen systems may be obtained with portable equipment, and this type of installation should be adequate for most small-aircraft requirements. Approved oxygen systems are available from a number of reputable manufacturers who offer detailed information for their different systems. If oxygen equipment is desired or planned for the near future, the design requirements of Ref. 13.11 plus the installation and service recommendations of Ref. 13.12 should be quite helpful. Again, in view of the extensive literature covering oxygen systems, it is not considered necessary to provide more detailed information in this chapter.

IFR EQUIPMENT

Equipment and installation reliability affect the level of IFR safety. Depending upon the amount and degree of IFR operation normally undertaken, autopilot and deicing systems may be required in addition to the desirable dual nav/com radio installations plus ADF, transponder, and DME equipment. To be useful, the autopilot must be maintained in adjustment, the gyros must operate consistently and not precess excessively, and all avionics equipment must perform with repeated accuracy; anything less represents a decrease in flight safety.

Such demanding requirements illustrate the equipment complexity accompanying sophisticated flight operation. Of course, initial purchase and routine maintenance costs also increase with equipment quantity. All of which is noted to underscore the importance of equipment to aircraft operation. For an airplane to be successful, equipment items cannot be left until last; their type, location, and accessibility must be considered along with the airframe as the design progresses. I trust this chapter has indicated how and why.

REFERENCES 13.1 FAR 23.397, 23.399, 23.405.

13.2 FAR 23.395 and 23.651(a).

13.3 FAR 23.561, 23.785, and 23.1413.

13.4 FAR 91.33, 23.1321, 23.1323, and 23.1325.

13.5 (a) AC 43.13-1A, pp. 303–310.

(b) AC 43.13-2, pp. 63–66.

13.6 FAR 23.1351–23.1401.

13.7 (a) AC 43.13-1A, pp. 173–212.

(b) AC 43.13-2, pp. 19–20-5 and 57–62.

(c) William J. Kendall, "Your Alternator or Generating System," *Business and Commercial Aviation*, May 1976, p. 60.

13.8 (a) AC 43.13-1A, pp. 291–301.

(b) AC 43.13-2, pp. 3–18-1.

(c) John Ferrara, *Every Pilot's Guide to Aviation Electronics.* El-Jac Publishing Co., 133 Glendale Drive, Trenton, N.J. 08618.

(d) Richard N. Aarons, "Antennas? Who Cares," *Business and Commercial Aviation*, June 1974, p. 52.

(e) Richard N. Aarons, "Planning an Add-on Radio Package," *Business and Commercial Aviation*, January 1976, p. 74.

(f) Richard G. Lawrence, "Catering to Your Black Boxes," *The AOPA Pilot*, January 1976, p. 44.

13.9 AC 43.13-2, pp. 63–66.

13.10 AC 43.13-1A, pp. 165–172.

13.11 FAR 23.1441–23.1449.

13.12 (a) AC 43.13-1A, pp. 154–158.

(b) AC 43.13-2, pp. 31–35.

(c) Hamilton Westfall, "Oxygen," *Aero Magazine*, April 1976, p. 16.

(d) D. H. Gollings, "Oxygen and Its Effects on Flight," *Air Facts*, May 1975, p. 6.

14 LANDING GEAR DESIGN

During the early forties aircraft design office arguments over the advantages of conventional vs. tricycle landing gear were only rivaled by the frequent arm waving and shouting considered necessary to protect buried liquid-cooled engines from those new, big, round air-cooled units. In each case the new development seems to have been the victor and sole survivor—until recently. We now find increasing numbers of aircraft fitted with conventional landing gear for special-purpose applications—such as Ag planes and acrobatic types—while time has replaced air-cooled radials with jet engines for high-performance operation.

TYPES OF LANDING GEAR
I have never seen a classification of the different types of landing gear built and flown over the years, although the many variations merit some recognition. Considering only the most frequently used, and therefore probably the most successful, we have:

1. Conventional landing gear, also known as a "tailwheel" or "tail dragger" type. The main landing gear wheels are located each side of the centerline *ahead* of the center of gravity, with a steerable tailwheel located aft near the rudder (originally a fixed tail skid was used and may still be seen on some antique aircraft). This type of landing gear was frequently noted for its ability to land "tail first," i.e., ground loop. Why this can happen will be explained a few pages further along.

2. Tricycle landing gear, known as the "nosewheel" type. The main wheels are located each side of centerline *behind* the center of gravity, with a free-swivel or steerable nosewheel mounted on centerline forward (just behind the propeller on most small landplanes). Usually a streamlined skid is fitted aft near the sternpost to prevent scraping the fuselage or tail surfaces during an extreme tail-down landing. This configuration is noted for ease of ground handling and will not ground loop unless steered into a skid. Originally developed by Glenn Curtiss back in 1908.

3. Four-wheel landing gear, two wheels mounted each side of the airplane centerline in tandem fashion. The main wheels are normally located behind the center of gravity, with smaller wheels at the forward position customarily occupied by a single (tricycle gear) nosewheel. This type of wheel arrangement is frequently used with amphibious floats, in which case the small wheel forward is mounted on the bow of the float. This is a very stable system if the forward wheels are free-swiveling.

4. Tandem-wheel landing gear, two wheels mounted one behind the other on the airplane centerline. The larger wheel is usually located just behind the center of gravity, with a smaller wheel forward in a nosewheel position. This arrangement has been carried to some sort of ultimate in the B-52 bomber design which has dual sets of tandem wheels retracting into the fuselage ahead and behind the bomb bay, plus outrigger wheels retracting into each outboard wing panel. Tandem gear is difficult to taxi in a crosswind, even with a steerable front wheel.

5. The single-wheel gear, usually mounted just forward or aft of the center of gravity with a landing skid to protect the fuselage. This type of gear is found on most sailplanes, with the wheel being manually retractable on higher-performance designs.

6. Skid gear, used by the Wright brothers for their early designs and then again through World War I, frequently incorporated into wheel landing gear systems for nose-over propeller protection. Skid gear is used on modern helicopters, and has been proposed for recoverable space vehicles.

7. Snow skis and floats may be used with any of the above systems.

It may be of interest to note that all of these landing gear arrangements, and many others, were designed and flown before World War I. Glenn Curtiss's 1908 June Bug was equipped with tricycle gear, a feature carried on all of his

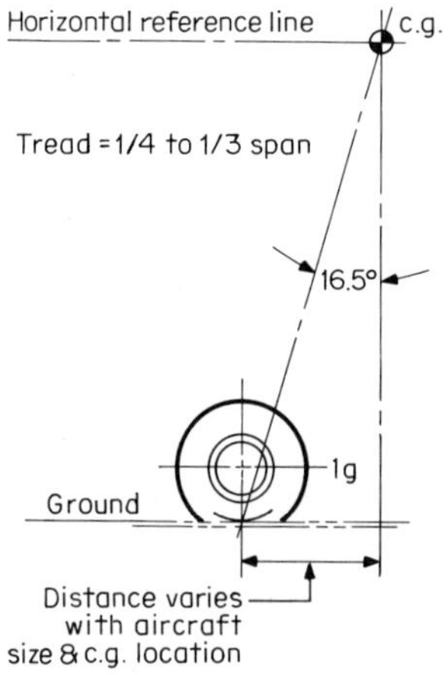

FIGURE 14-1
Conventional gear. Location of main wheel with respect to center of gravity.

Horizontal reference line — c.g.

Tread = 1/4 to 1/3 span

16.5°

Ground — 1g

Distance varies with aircraft size & c.g. location

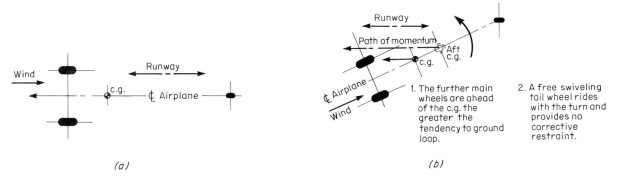

(a)

(b)

Runway

Wind

c.g.

℄ Airplane

Runway

Path of momentum

c.g.

Aft c.g.

1. The further main wheels are ahead of the c.g. the greater the tendency to ground loop.

2. A free swiveling tail wheel rides with the turn and provides no corrective restraint.

℄ Airplane

Wind

FIGURE **14-2**
Stability characteristics of conventional landing gear. (*a*) Conventional gear tracking down runway. (*b*) The ground loop starts. If the momentum vector gets outside the tread, a ground loop is underway. Note that the aft center of gravity requires a smaller skid angle to start the windup. Tracking correction must be made immediately.

designs through 1913 and adopted by the Wright brothers, Breguet, and Cody during 1910.

Dual-wheel conventional-type gear was also very popular with early designers. Practically all aircraft types used this gear at some time and with minor variations. Since available bicycle wheels could not support aircraft weights at high speeds on rough ground, two bicycle wheels were frequently mounted side by side on the same axle, usually with a protective skid between to save the propeller and engine during nose-over accidents.

Four-wheel gear was developed by AVRO in 1907 and was subsequently used by Sopwith, Bristol, M. Farman, H. Farman, and Cody among others. Tandem-wheel gear was first introduced on the 1907–1910 series of REP aircraft and has reappeared over the years for special applications, but remains the least popular of surviving landing gear configurations.

So much for early landing gear history; let us now examine the relative merits of conventional and tricycle landing gear from the design and operational standpoints.

CONVENTIONAL GEAR

Conventional landing gear design normally places the main wheels on a line running about 16.5° ahead of the center of gravity, with the airplane in a level attitude per Fig. 14-1. This position is based upon a study of 10 aircraft, including twin-engine amphibians. For aircraft having a high thrust line, it will be necessary to increase the angle by 4 or 5° to prevent nose-over during engine runup and rough-field operation. However, the further forward the main wheels are placed, the larger the tailwheel loading. And the heavier the tail load, the greater the tendency to ground-loop, as may be noted from Fig. 14-2.

A free-swiveling tailwheel is likely to ground-loop during crosswind operation because it is free to turn with the tail end of the airplane, and so offers no resistance to rotational momentum once the tail starts to swing. Fortunately, most tailwheel-type aircraft are equipped with a steerable wheel that remains engaged until the rudder has been turned about 30° off the airplane centerline. This provides lateral resistance to sidewise motion of the tail. Unfortunately,

30° may be the amount of rudder deflection suddenly applied to correct a severe skid condition. And then, just when tail restraint is most required, the tail wheel releases into free-swivel and away you go upwind toward the woods or runway lights as shown by Fig. 14-2b.

This embarrassment can be reduced, if not eliminated, by use of a locking tailwheel. This system physically locks the tailwheel to the fuselage or tail spring structure by a cockpit-controlled pin actuated during the landing approach check. Before World War II, when conventional gear planes were practically the only type built, all carrier-landing aircraft used locking tailwheels, as did most of our military and transport designs. But since it cost something to install and was one more complication for the pilot to unravel (for example, it is most difficult to taxi around corners with the tailwheel still locked), the locking tailwheel never became popular on personal aircraft.

If conventional landing gear can be a bit much to handle in crosswind conditions, why does it persist? Principally because this arrangement is excellent for

1. *Rough-field and short-field operation.* The tail may be raised clear of terrain early in the takeoff run and trim attitude set for maximum wing lift as discussed in Chap. 6. If it were possible to raise a tricycle gear nosewheel clear of the ground at 20 mph, so much elevator deflection would be required that the accompanying drag would prolong or prevent takeoff (see Fig. 6-1).

2. *Amphibian operation.* Without a nosewheel it is possible to approach a ramp in cross current and/or crosswind conditions without first grounding on the nosewheel and pivoting about to lie across the ramp. With conventional gear both main wheels touch the ramp at the same time, permitting direct taxi up and out onto the parking area. And when beaching, application of power lifts a tailwheel clear but buries a nosewheel in the sand.

These features are accompanied by lighter installed weight, cheaper manufacturing cost, and reduced maintenance compared to a tricycle gear. As a result, the conventional gear is particularly well suited to agricultural and sport aircraft and is extensively used on these types today.

Those of us who fly for pleasure or occasional business purposes should be alert when taking off or landing crosswind with conventional gear. But if you keep ahead of the airplane, practice will develop confidence. Because of main gear location, tailwheel load, and tread some conventional gear aircraft handle easier than others; for example, I know my PT-19 handled much better than any PT-17 or Luscombe 8 ever did during landing runout. And grass airfields were much easier to operate from than paved runways.

To relieve the crosswind landing problem, development of automatic or semiautomatic runway alignment landing gear flourished during the mid- and late forties. Generally known as "crosswind" or "self-castering" gear, various designs were developed by J. Geisse, Goodyear Aircraft, and B. W. King among others. The principle of operation is shown in Fig. 14-3; the main wheels were

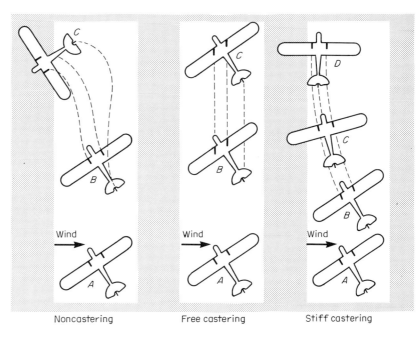

Wind   Wind   Wind

Noncastering   Free castering   Stiff castering

FIGURE 14-3
The principle of self-castering
main gear

arranged to caster and trail along the path of momentum straight down the runway. They occasionally would also caster without warning when taxiing around corners, were heavier and more expensive than the plain wheel and axle combination, and required more service adjustment and attention. These disadvantages, combined with the arrival of the new generation of postwar tricycle gear aircraft, precluded full development and acceptance of the crosswind gear concept—which, incidentally, Louis Bleriot had first incorporated on his English Channel crossing airplane back in 1909. A version of the old castering main wheel is still with us on the C-5 military transport. This huge airplane can steer its main gear up to 18° from the fuselage heading to land and run out in a crabbed attitude. Heaven help us all if the pilot should forget just before touchdown and kick out the crab with the main wheels cocked 18°.

In summation, conventional-type landing gear is superior for rough-field aircraft and working amphibians, but requires more pilot skill and technique, particularly during crosswind operation. In addition, a conventional gear installation is lighter, cheaper, and requires less maintenance than a corresponding tricycle gear. So it is an ideal landing gear for a working airplane and professional pilots.

TRICYCLE GEAR There certainly can be no doubt that tricycle landing gear has been a major factor in the expanding popularity and use of modern aircraft. Approach stability, longitudinal trim, and tricycle gear combine to make landing almost auto-

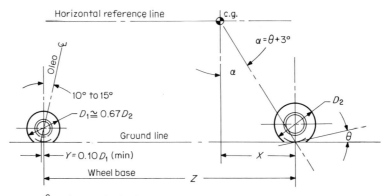

θ = tail angle for 0.90 C_L max, flaps down

FIGURE **14-4**
Tricycle gear arrangement

X = 0.09Z to 0.15Z

Tread = 0.75Z to 1.25Z

matic, with ground handling no longer a major problem for the occasional pilot.

Figure 14-4 presents an average configuration for an acceptable tricycle landing gear design. Note that the tail clearance angle is based upon the angle of attack for 90 percent of the maximum lift coefficient, since approach is always at some speed above stall. Ideally, the tail bumper or skeg should just clear the ground at this angle, or an angle 1 or 2° greater. The main wheels should be positioned behind the most aft empty weight center of gravity so that the center of gravity always remains ahead of the wheel contact point; if the main wheels are too far forward the airplane will squat on its tail with the center of gravity *behind* the main gear. With these basic technical aspects of tricycle gear design in mind, let us review the practical benefits and problems.

Why does the tricycle configuration perform so well on the runway? As described in Fig. 14-5, the location of the main wheels behind and outboard of the center of gravity makes them tend to trail directly behind the center of gravity. This action is called *inherent dynamic stability*. We all know it is easier to pull a two-wheeled cart uphill than push it, particularly if the tow bar is pinned to the cart. And pushing a cart with a pinned bar is similar to the action of a conventional gear with free-swiveling tail wheel.

Despite the rather obvious (when diagrammed) stability of the tricycle arrangement, a number of operational problems had to be resolved before tricycle gear became practical. Among those were the severe tendency for nose-wheel shimmy, the proper distribution of weight between the nose and main gear, the correct relationship between wing lift and main gear location, adequate tail power to rotate about the main gear at relatively low speeds, and reduction of installed landing gear weight.

All these factors do not lend themselves to ideal solutions; for example, it is impossible to obtain short-field takeoff performance from tricycle gear aircraft that will be superior to conventional gear designs of similar weight and wing area, a condition particularly apparent during rough-field operation as previously noted. The Helio and Maule designs retain conventional gear for this reason, as do the Cessna 180 and 185, plus the fact that tricycle gear would have greater installed weight and cost.

Another problem inherent in the tricycle gear arrangement is shown in Fig. 14-5c, resulting from misuse of the steerable nosewheel. In fact, if the main gear or nosewheel doesn't collapse first, the condition shown will produce a ground loop as severe as any experienced with the most savage conventional gear. This problem used to occur in varying degrees with the two-control Ercoupe, which had interconnected rudder and ailerons plus a nosewheel steered by the control wheel for ground operation. Crosswind work had to be an approach banked upwind and carried well into the flare, with the nosewheel also turned upwind through the interconnected control system. If one didn't level out at just the proper moment, the plane either skidded downwind (recovery too early) or went smartly around upwind (recovery too late and the nosewheel still pointed upwind). As a safe alternative, the Ercoupe could be landed on the main gear as crabbed and the nosewheel straightened before it contacted the ground.

Fortunately, the final approach procedure with three-control tricycle gear aircraft will normally be a bank into the wind with some top rudder; so even if the nosewheel and rudder steering are interconnected, runway contact will usually occur with the nosewheel aligned with the direction of motion—or at least that should be the intent. This also explains why modern nosewheels have spring bungee steering which allows the drag load to straighten any nosewheel offset angle present at runway contact.

Since most of today's flying is from paved runways, and since most of us prefer to be relaxed and humming away during final rather than "white knuckling" our prospects for ground-looping a severe crosswind landing, we should not be surprised that tricycle gear aircraft dominate general aviation use. But conventional gear designs will see continued service because this arrangement is superior for certain applications.

Whether tricycle gear or conventional, the use of wheel fairings will improve

FIGURE **14-5**
Stability characteristics of tricycle landing gear. (a) Tracking down runway. (b) Recovery with a free-swiveling nosewheel. Momentum acts to recover from a skid. (c) Windup possible with a steerable nosewheel when holding rudder at touchdown to correct for sudden crosswind gust.

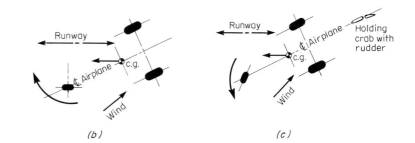

(a) (b) (c)

climb and cruise performance as shown by Fig. 6-10. In fact, the resulting climb performance halfway approaches that of fully retracted landing gear, but without as much maintenance, cost, and weight. Unfortunately, winter operation and wheel fairings are not particularly compatible in northern areas because of frozen slush and locked wheels, so fairings are usually removed before the first snow flies.

SKIS As an alternative, skis may be installed per Ref. 14.1. Most ski planes are of conventional gear type primarily due to the short-field performance offered. In addition, there is some skepticism about the use of skis on tricycle gear aircraft, although this arises largely through inadequate nosewheel design loads for snow operation rather than from any ground handling problems. The nosewheel strut and attachment structure are frequently much too weak to withstand frozen snow drift and ice ridge contact at high speed.

Tricycle gear aircraft intended for snow operation should be designed with all nosewheel strut and support structure load factors increased to twice the required landplane values. Although this will add somewhat to the airframe weight and cost, it is nothing compared to the inconvenience of being stranded in bush country at 20° below zero with a torn-out nosewheel and damaged prop.

ENERGY-ABSORBING SYSTEMS A brief look at the popular types of energy-absorbing systems will provide some practical background and useful design information. Also known as "shock struts" since they must withstand severe impacts, landing gear struts are designed to absorb dynamic loads by using one of the following energy absorption control methods:

a. Shock cord

b. Rubber discs

c. Air and oil (also called an oleo strut)

d. Fluid alone

e. Trailing beam with one of (a) through (d) shock absorbers

f. Springs, which may be simple leaf spring struts or compression spring/oil types

With the exception of the air and oil (or oleo) strut all the methods really dissipate very little energy by themselves, and what is absorbed is removed in the form of friction heat. Referring to Fig. 14-6 for each description:

a. Shock cord is a simple and cheap way of controlling wheel deflection to ab-

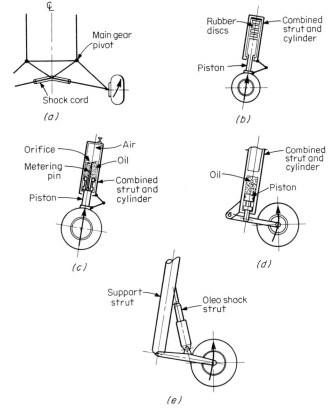

FIGURE **14-6**
Schematic drawings of various types of landing gear energy-absorbing systems. (a) Shock chord. (b) Rubber discs. (c) Oleo strut. (d) Liquid spring. (e) Trailing beam strut.

sorb vertical velocity. But except for friction developed between elastic components, no impact energy is absorbed by shock cord. Of course, axle displacement permitted by the shock cord plus tire deflection reduce vertical impact velocity. This in turn reduces the landing gear load factor and airframe loading, as shown in the spring strut design example of Ref. 14.2.

b. Rubber discs act quite like shock cord, dissipating some energy by generating internal friction and possibly surface friction as well through contact with a cylinder wall. Again, the resulting wheel and tire deflections combine to reduce gear impact loads imposed upon the airplane structure.

c. Air and oil can be used to provide a desired rate of strut deflection by controlling flow through an orifice. The air chamber above the oil in Fig. 14-6c is pumped up to a pressure sufficient to support the static load on the strut when the airplane is at rest. Impact loads experienced during taxiing over rough ground or when landing will compress the air and so increase strut deflection, forcing oil up through the orifice. This action dissipates (or absorbs) some energy while also controlling the rate of strut deflection, and so reduces landing loads by eliminating shock peaks. A low sink rate at wheel contact will require only a small deflection of the oleo strut to accept the landing load, which would be similar to the static loading; while a high sink rate at ground contact will cause large strut deflections at a rate controlled by the orifice design. To provide the low impact load factor desired, orifice characteristics must be developed and proven by drop tests.

The metering pin shown may not be necessary to regulate orifice area vs. displacement for very light airplanes since a simple drilled hole may be adequate, but it is mandatory for heavier models due to the great variations in landing loads possible with larger aircraft.

The oleo strut is an ideal method of controlling impact loads, as evidenced by its use on carrier-based and transport aircraft. Since an oleo strut requires more maintenance and is more expensive than simple spring leg struts, it is not as popular for small private aircraft, even though an oleo main gear should be lighter than a leaf spring steel strut.

d. Although oil is considered to be a virtually incompressible fluid, Dowty developed a liquid spring strut many years ago based upon the trailing beam design of Fig. 14-6d and e. As fluid pressure increases, the containing cylinder deflects, permitting the liquid strut to compress and the wheel to move up. However, the high pressures involved preclude use of this type of absorber on small aircraft because of the high cost and weight.

e. The trailing beam strut may have shock cord, rubber discs, fluid, or an oleo strut for wheel displacement control as shown in Fig. 14-6e. The great advantage of this type of strut lies in its ability to smooth out rough terrain and equally rough landings. It also is a nice design for amphibian aircraft because it permits brake application on a wet ramp without causing main strut binding and resulting fore and aft chatter.

f. The combination of oil, a compression spring instead of air, and an orifice used to be quite popular. This approach eliminated the need to monitor air pressure for correct operation, but the cylinder had to be fluid tight nonetheless. Unfortunately, coil compression springs tend to set with time and would frequently bottom with a bang if the oil level became too low—as it frequently did on the PT-19 which had this type of main strut. So progress has favored development of the leaf spring gear, whether it be of fiberglass, aluminum, or steel.

Spring struts are very much with us today. Introduced by Steve Wittman and further developed by Cessna Aircraft, the leaf spring main gear strut of Fig. 1-3 is probably flying on more aircraft than any other type of landing gear. While leaf springs do not provide the energy control of an air/oil orifice system, they have the distinct advantages of extreme simplicity and virtually maintenance-free operation. Steel is the most effective material for leaf spring landing gear legs because of its strength characteristics. Since this type of strut is of interest to many owners and designers, an example of steel strut design is presented as Ref. 14.2. These calculations show how root section dimensions affect strut energy characteristics and so can vary the load factor used for landing gear and attachment structure design.

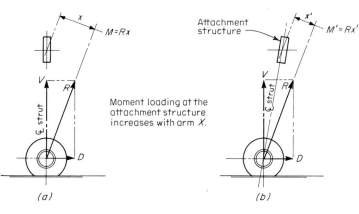

FIGURE **14-7**
Main gear load reactions.
(a) Vertical strut. (b) Inclined
strut.

As a final comment regarding energy absorption, it is advisable to slant main gear struts slightly forward and down to align more closely with the combined vertical and drag load resultant as shown in Fig. 14-7. Whether of conventional or tricycle gear type, you have probably noticed that most main struts incline forward. By doing this the designer has decreased bending loads on the strut and attachment fittings while also reducing the wearing loads on oleo struts. Although trailing beam shock absorbers are always axially loaded and so not subject to bending loads and friction, their attachment struts will be placed in bending by landing and brake loads; so even trailing beam gear legs should be inclined forward to reduce support strut and attachment bending loads.

**LANDING GEAR
DESIGN REQUIREMENTS**

The FAA design conditions listed in Ref. 14.3 require each main wheel to be at least capable of supporting the airplane gross weight, thereby establishing the tire load rating and size. That load must, of course, be multiplied by the 1.50 ultimate load factor to obtain the actual main gear design numbers. This means a 2400-lb gross weight airplane, whether of conventional or tricycle gear configuration, must be able to carry a load of 2400 × 1.50 = 3600 lb through each wheel and landing gear strut into the airframe attachment structure. And sometimes even this is not enough strength for student landings; so trainers may have even larger load factors for landing gear design.

Nosewheels have to withstand about one-third the airplane weight both vertically and in an aft direction, with tailwheels subject to similar loadings. The horizontal loads anticipate obstructions such as bumps and furrows, but are really not adequate for sustained rough-field operation or nose-ski installations as previously noted. Both nosewheel and tailwheel design factors of Ref. 14.3 should be doubled for continuous heavy-duty use.

To prove just how rugged the final landing gear design really is, as well as to substantiate the load conditions calculated for the main gear and nosewheel or tailwheel, FAA tests are again necessary. The specifications listed in Ref.

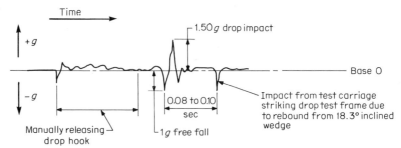

FIGURE **14-8**
A typical landing gear drop test record

14.4 require a series of controlled drop tests, including one from 1½ to 2 ft above the ground. A loaded airplane may be used for these tests, or the individual main strut and tail or nosewheel units can be tested separately.

Either way, the strut being tested must be loaded to limit design conditions and dropped from increasing heights until the required distance has been reached—hopefully without breaking the wheel, strut, or any of the attaching bolts. The impact vs. time history of each drop is recorded and then measured to determine the actual load factor as shown in Fig. 14-8. Most of the main gear drop tests will include impact onto an inclined wedge to simulate a fully braked or wheels-frozen landing condition, producing not only severe torsional loading of the main gear attachment structure but also some side effects in the data picture as noted on Fig. 14-8.

Despite the fact that all landing gear components plus the attachment structure have been thoroughly tested and are capable of taking the expected level of normal abuse, it is still advisable to check regularly the various wheels, struts, and structural connections to be sure nothing has worn excessively, worked loose, or started to crack out around bolts and bushings. The landing gear is probably subject to the most fatigue and impact loading of any part of the airplane, and should receive sufficient attention to ensure it doesn't literally let you down—embarrassing at best, as well as being very hard on the propeller, engine, wings, lower fuselage, and pocketbook.

BRAKING SYSTEMS   Modern aircraft are fitted with hydraulic disc brakes because this type has less tendency to fade than drum brakes and will immediately wipe water, slush, and mud from the braking surface. A steel brake disc is fitted to the inside face of the cast magnesium or aluminum wheel, while the friction-producing brake lining material is fitted onto the facing of small pistons. Hydraulic pressure moves these pistons to clamp the brake disc in a caliperlike action whenever the brake master cylinder is operated in the cabin.

Wheel braking action may be applied (1) to both wheels at the same time by a single lever as used on Cherokees; (2) to either or both wheels by toe brakes actuated through rudder pedal deflection; and (3) independently or together by separate heel brakes. Toe brakes may be located only on the pilot side or at both front seat positions, while heel brakes are used in tandem arrangement on the Piper SuperCub; so we have a multiple choice of braking systems.

Fortunately, toe pedal action is rapidly becoming the standard method of brake application, with dual installations used to increase operational safety for instructional and demonstration aircraft. The single brake lever is fine when parking or to stop overrunning a taxiway or runway, but it offers no assistance when turning sharp corners or for independent wheel braking action frequently needed in high-wind conditions. Heel brakes are clumsy and difficult for the occasional pilot to operate, and they can be missed when heels start moving fast for urgently required braking action (to prevent a ground loop, for example).

In view of these deficiencies it is not surprising that as the number of private pilots increased, so did the demand for independent toe brakes. Piper has combined braking systems to offer both individual toe brakes and the single emergency/parking brake lever on the Cherokee series; so whether you are used to pulling a single lever or depressing toe pedals you can be sure of stopping in a hurry.

Parking brakes are now standard equipment on most modern aircraft and may be applied (1) by depressing and locking the toe brake master cylinders on the pilot's side; (2) by depressing toe or heel pedals and locking the brake pressure in the lines by actuating a hydraulic locking valve; or (3) by pulling back and locking a single brake lever.

Either way, a control on or under the instrument panel will be pulled to engage the locking action, while the brakes will be released either by (1) releasing the locking lever; (2) depressing the toe or heel pedals; or (3) performing both actions at once. The problem with all parking being that hydraulic pressure is trapped in the brake lines to energize and lock the wheel brake pads onto each brake disc. If ambient temperature increases sharply after applying parking brakes, the hydraulic line pressure may also increase to a level capable of bursting the flex line or some other weak connection in the system.

To prevent such failures, hydraulic brake systems must be tested to 1½ times the line pressure possible from the 200 lb maximum force a pilot can apply onto the brake pedals (Ref. 14.5). But even this test does not assure freedom from ruptured lines at high temperatures, particularly in older aircraft. So if the day is going to be a hot one, check your wheels and do not use the parking brake.

RETRACTION SYSTEMS

While wheel fairings will increase climb rate and cruising speed, high-speed landplanes must retract their landing gear to realize maximum performance. Of course, all amphibians must retract their gear prior to water operation.

Manual retraction is probably the simplest system and one requiring minimum maintenance. The actual retraction may be accomplished by direct hand lever action or by use of a hydraulic hand pump. The latter method has considerable merit for smaller aircraft if hydraulic pressure is stored in an accumulator until released to raise or lower the gear by actuating a selector valve.

Manual systems have another distinct advantage in that alternate retraction methods are not required by FAA regulations. If motor-driven hydraulic pump pressure or electrical retraction drives are used it is necessary to supply a backup auxiliary hand pump or hand-crank method of lowering the gear in case the primary system fails. With hand power applied through a lever mechanism or pump handle we have only the basics, and this fact is recognized in FAA specifications.

Whenever the throttle is pulled back below normal base leg or approach power settings, land aircraft have lights and buzzers to warn that the gear is up. Some aircraft also have automatic gear extension systems that lower the gear whenever flight speeds get below normal pattern levels. This system may be manually cut out for low-speed flight practice, but will be quite useless unless reactivated prior to entering final.

Amphibians do not have such warning systems for the simple reason that the normal reaction to a flashing red light and horn signal would be to extend the gear—during a water approach. As a result the number of amphibian landings on land with the gear up is sufficiently large to distort amphibian hull insurance rates, while the number of landings on water with the gear down is also distressingly large. The best answer for all aircraft operations is to establish a complete checklist for each phase of flight and then use it constantly, especially when flying an amphibian.

You have probably noticed that all aircraft usually retract the main gear inboard toward the wing root. This is done because outboard retraction places the gear weight further out in the wings, which in turn increases the airplane lateral mass moment, also known as *lateral inertia*. Increasing the lateral inertia results in either the installation of larger and more powerful ailerons to maintain a specified rate of roll, or acceptance of a reduced rolling rate for an established wing configuration. So for new designs and when converting an existing model from fixed to retractable gear, the main wheels are folded inboard whenever possible.

$$* \qquad * \qquad *$$

This chapter completes our landplane design review. I hope the next chapter about seaplane design will be of general interest to most readers, since amphibian utility and safety have so much to offer all sport and business pilots. Following seaplane design, chapters covering the mysteries of aircraft protective finishes and FAA certification procedures have been included to complete every reader's background from basic aircraft design through final approval.

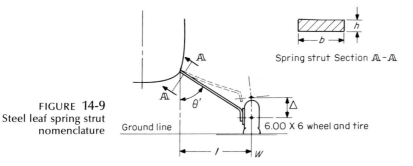

Spring strut Section A-A

FIGURE 14-9
Steel leaf spring strut
nomenclature

Ground line

6.00 X 6 wheel and tire

NOTES AND REFERENCES

14.1 (a) David B. Thurston, "North Country Skis," *Aero Magazine,* November/December 1975.

(b) AC 43.13-2, pp. 21–29.

14.2 Sample design calculations for an acceptable steel leaf spring main gear strut:

(a) FAA design conditions require *each main wheel* to carry a vertical load at least equal to the airplane gross weight per FAR 23.473(g) and FAR 23, Appendix C. This is equivalent to a 2 *g* impact load on the main wheel and strut. The following deflection formula will establish acceptable strut cross-section dimensions based upon calculations developed for the Teal amphibian and later substantiated by landing gear drop tests.

(b) Based upon the accompanying sketch we have per Fig. 14-9:

$$\Delta = \text{strut deflection at the axle under 2 } g \text{ impact load}$$

$$= \frac{Wl^3}{3EI} \times \text{Sec } \theta', \text{ in inches}$$

where W = 2200 lb for a 2200-lb gross airplane

l = 22.5 in. for this design

E = the modulus of elasticity = 30×10^6

for steel (10×10^6 for aluminum alloy)

where I = moment of inertia of the strut root section = $b(h)^3/12$, in (in.)^4

θ' = 58° for this design

(c) Looking at the formula it is quite obvious that I holds the key to adequate deflection of the strut under impact load conditions. If I is too large, the deflection will be too low and the impact load factor will be high; in other words, the gear will be too stiff. The following calculation shows the effect of deflection upon the design load factor. When developing a strut design it will be necessary to vary strut dimensions b and h until sufficient deflection is obtained to provide acceptable vertical load factors—and hopefully so without a too soft or "springy" gear.

(d) If the strut selected is 3 in. wide at the base (b = 3) and 0.69 thick (h = 0.69), with θ = 58°:

$$I = \frac{b(h)^3}{12} = \frac{3 \times (0.69)^3}{12} = 0.082 \text{ (per section A-A above)}$$

Secant $\theta' =$ secant $58° - 1.89$

$$\Delta = \frac{2200 \times (22.5)^3}{3(30 \times 10^6)(0.082)} \times 1.89 = 6.41 \text{ in. strut deflection}$$
at the axle at 2 g impact.

(e) From 6.00×6 tire deflection charts at an inflation pressure of 20 psi, tire deflection $= 3.14$ in. at 2200 lb loading.

(f) Therefore, the total axle travel to descent velocity $V_d = 0$ will be: 6.41 strut deflection $+ 3.14$ tire deflection $= 9.55$ in. $= 0.796$ ft stopping distance $= d$.

(g) According to FAR 23.473(d), the initial descent velocity for landing gear design calculations may not be less than $4.4(W/S)^{1/4}$, where $W =$ gross weight and $S =$ wing area. Assuming $S = 157$ sq ft, we have $V_{dmin} = 4.4(2200/157)^{1/4} = 8.5$ ft/sec. To find the landing gear vertical g design inertia factor: $(V_d)^2/2d = (8.5)^2/(2)(0.796) = 45.4$ ft/sec² and the vertical design inertia factor $= 45.4/32.2 = 1.4$ g due to stopping descent in 0.796 ft.

(h) The inertia load factor, $\eta = 1.0 + 1.4 = 2.4$ g (since the airplane and gear were subject to 1 g load in steady flight before the additional impact of 1.4 g was added at landing).

(i) Since FAR 23.473(e) permits a load reduction of 0.67 g due to wing lift assumed to act during landing, the landing gear design load factor becomes $2.40 - 0.67 = 1.73$ g, indicating acceptable strut deflection.

(j) Because FAR 23.473(g) requires at least a 2 g limit design load factor, the landing gear limit design load factor used must be 2 g and not 1.73 g as indicated.

14.3 FAR 23.471–23.509 and Appendix C.

14.4 FAR 23.723–23.727.

14.5 FAR 23.1435(b).

15 SEAPLANE DESIGN

A *seaplane* is an airplane capable of operating off water. Seaplanes may have twin floats or be designed with a single main float plus small wing floats or similar lateral supports. The single-main-float seaplane is frequently referred to as a flying boat. And when any seaplane is equipped with wheels it becomes an amphibian, an airplane capable of operating from land or water.

Various seaplane float, hull, and wheel configurations have been illustrated and discussed in Chap. 4. So this review will primarily cover seaplane design requirements, followed by some practical suggestions for successful and safe water operation that should be of interest to anyone flying off water.

Unfortunately, due to the few fixed-base operators catering to water aircraft, the pure seaplane is not as popular as it should be. For example, it is difficult for seaplanes to refuel when flying cross-country unless special plans are made for advance service, or unless the seaplane is an amphibian capable of land operation.

Since amphibians combine land, water, and flight operational problems in one configuration, we will concentrate upon that type. For seaplane design where only water operation is desired, simply remove the landing gear, thereby increasing useful load while also reducing purchase and maintenance costs.

THE AMPHIBIAN —A SAFE UTILITY AIRPLANE

The primary reason for the existence of amphibian-type aircraft may be summed in one word—*utility*. A very close secondary reason is *safety*.

The amphibian is designed to operate from rough airports and available waterways, frequently without any prepared surface, docking, or handling facilities. As a result it is a most effective form of transportation into remote and undeveloped areas, where amphibian use will increase as emerging countries realize their economic potential. Fuel capacity for 6 to 10 hr of cruise operation, coupled with landing versatility, permit the amphibian to fuel at home base, cruise outbound for 3 hr or more to a remote camp or outpost, and return home without refueling. While in transit, every river, pond, or lake over ½ mi

227

long is an emergency landing field, as are the usual open fields and intermediate airports. No other aircraft type encompasses such a broad spectrum of utility—3 to 5 hr cruising radius plus the ability to operate from water or land.

In addition, the rugged hull bottom resulting from water loads offers added protection in the event of a forced landing away from water or a suitable field, and this is a real safety item of benefit to every owner.

All of these features combine to offer a safe type of airplane for sports pilots as well. If they wish to reach a remote lakeside camp, amphibian pilots can fly from their home airport and land on the water at their dock. And when ski-equipped, a wintertime campsite landing may be made on the ice. Such capability provides year-round entry into inaccessible areas, and in all climates.

If the amphibian airplane is so versatile and capable, why isn't the type more popular? Because as usual there are several penalties accompanying the numerous advantages. The initial cost and weight of an amphibian compared with a landplane of similar capacity is greater because of the rugged hull and larger powerplant required. Secondly, operating costs are somewhat higher because of lower cruising speeds resulting from the hull shape, wing floats, engine location, and possibly the landing gear hanging out into the airstream. Useful load capability may be traded off against power vs. climb requirements, so an amphibian might not carry as much as is occasionally desired, especially on a hot day. And finally, some added maintenance will result from the ability of water to find its way into almost every part of the airframe under the high pressures and velocities associated with water flight.

It is obvious that the amphibian is a compromise design combining water, land, and air operation into an acceptable vehicle. This is a most challenging problem, with the really unattainable ultimate being a configuration economically competitive with a landplane of similar power and gross weight. While water handling, power, and flotation design requirements preclude full realization of this goal, it can be nearly achieved through use of a surface-piercing hydrofoil positioned on a faired step or stepless hull as will be discussed under hull design.

NOMENCLATURE
: The following nomenclature applies to seaplane design and related formulas used in this chapter:

b = maximum beam at the chine, ft

c.b. = center of buoyancy

C_{Δ_o} = gross load coefficient = Δ_o/wb^3, dimensionless

g = acceleration of gravity

H = step depth, ft

K = spray coefficient = Δ_o/wbL_f^2, dimensionless

L_a = length of afterbody, ft

L_f = length of forebody, ft

MAC = mean aerodynamic chord

n_{w_s} = water landing impact load factor at step

$P_{H.T.}$ = horizontal tail power coefficient, dimensionless

$P_{V.T.}$ = vertical tail power coefficient, dimensionless

S_H = hydrofoil area, sq ft

$S_{H.T.}$ = horizontal tail area, sq ft

$S_{V.T.}$ = vertical tail area, sq ft

S_W = wing area, sq ft

V = velocity, knots

V_{so} = stalling speed in landing configuration, knots

w = specific wt. of water, lb/ft^3: 62.4 /ft^3 freshwater; 64 /ft^3 seawater

W = gross weight, lb

y = distance from center line to float c.b., ft

α = sternpost angle, degrees (see Fig. 15-1)

β = angle of deadrise, keel to chine, degrees (see Fig. 15-1)

β_k = angle of deadrise at keel, degrees

Δ_o = gross displacement lb

Δ_{tf} = tip float displacement (freshwater), lb

τ = trim angle, degrees

HULL DESIGN

Although major components of any airplane are essential to its performance, hull design and resulting water handling characteristics are particularly critical for an amphibian since it exists only because of its capability for water operation. In this regard, it should be noted that we have now reached a point in seaplane development permitting hull forms to be classified as conventional (having a transverse step) or advanced (either stepless or with a faired step). Each type will be discussed separately in the following paragraphs.

1. *Conventional Hull Design* The classic hull shape shown by Fig. 15-1 may be referred to as a planing tail hull since it only has one step with a forebody ahead of the step and an afterbody behind the step, each with V-shaped bottom surfaces for reducing water impact loads. Wing-mounted floats provide lateral stability when on the water. Conventional hull shapes and performance char-

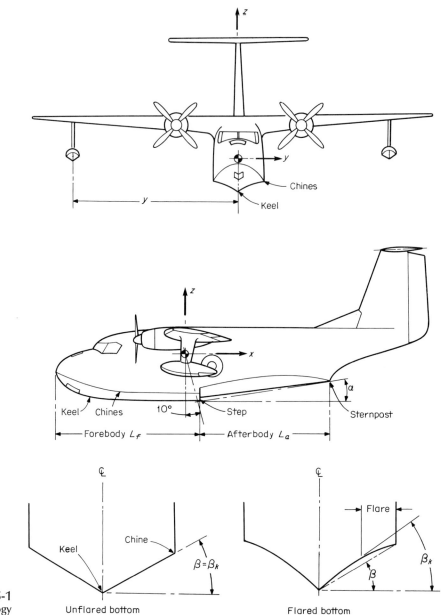

FIGURE **15-1**
Hull terminology

acteristics for small flying boats are thoroughly evaluated and well presented in NACA TN 2503, Ref. 15.1. The trim advantages of slightly inclining the forebody keel 1 or 2° up and forward are apparent from TN 2503 data presented for Model 1057-04.

Forebody hull bottom deadrise angle β should not be less than 15° for the smallest hull and gross weight, increasing to 25° at the step for larger seaplanes in order to reduce design water load factors as shown by Fig. 15-2. Exceeding

20° deadrise, however, will create ground clearance problems for most configurations. As the bow is approached, deadrise may be increased to 40° by simply warping the bottom sheet on small hulls, while larger seaplanes will benefit
from reverse flare at the bow—although this will require stretched skins and is
costly. In either case, increased deadrise at the bow reduces wave impact
loading and increases sea-state capability for a given forebody length and
beam.

Figure 15-2 also dramatically illustrates the important effect of V_{so} upon hull
bottom loading. The design water load conditions are set forth in FAR Part
25.527(a) (1) and subsequent paragraphs for all commercial seaplanes, with
Part 23 aircraft also covered in that section. The basic water loads formula is
empirical and has been developed from operational experience:

$$n_{w_s} = \frac{0.012\,V_{so}^{2}}{\sqrt[3]{(\tan^{2}\beta)\,W}}$$

where n_{w_s} = water landing impact load factor at the step

$\quad V_{so}$ = stalling speed in knots (for all FAA water loads)

$\quad \beta$ = forebody deadrise angle, degrees

$\quad W$ = gross weight, lb

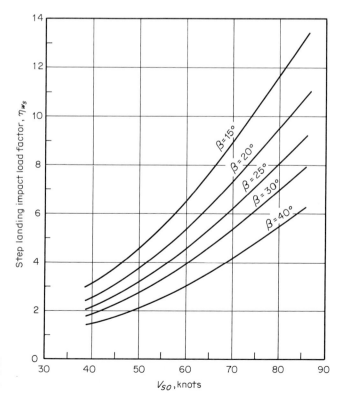

FIGURE **15-2**
Step landing impact load
factor variation with landing
speed and deadrise angle
for 4000-lb seaplane

Since water loads are a dynamic impact function increasing with velocity squared, lowering landing speed by 30 percent will reduce the bottom loading nearly 50 percent as shown by Fig. 15-2.

The values selected for deadrise angle and landing speed are critical considerations for hull design because they affect the basic hull weight and cost as well as the sea-state capability. And a heavy hull is not only expensive, it also reduces useful load.

The basic relationship between forebody length, maximum beam, and beam loading is approximated by Fig. 15-3 which applies to floats and hulls. When calculating C_{Δ_o} and K values for seaplanes under 5000 lb displacement and fitted with spray rails, the beam value b may be extended to cover the width over spray rails at the step. Figure 15-3 data show a reduction in spray with in-

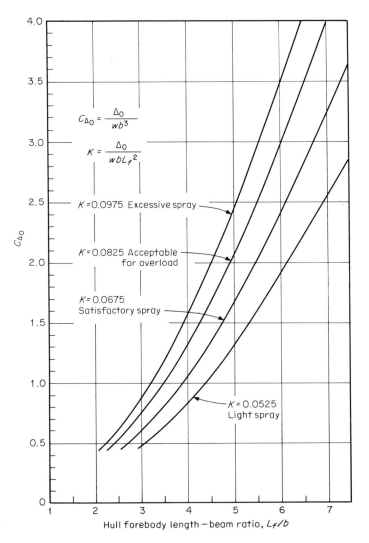

$$C_{\Delta_0} = \frac{\Delta_0}{wb^3}$$

$$K = \frac{\Delta_0}{wbL_f{}^2}$$

$K = 0.0975$ Excessive spray

$K = 0.0825$ Acceptable for overload

$K = 0.0675$ Satisfactory spray

$K = 0.0525$ Light spray

Hull forebody length – beam ratio, L_f/b

FIGURE 15-3
Forebody length and beam
vs. hull spray characteristics

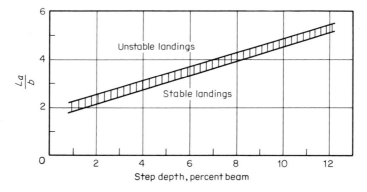

FIGURE **15-4**
Afterbody length and beam
vs. step depth

creased forebody length; and since spray is a direct indication of hull drag, there is a natural tendency to extend the forebody. However, any excess vertical area ahead of the center of gravity further reduces the marginal directional stability inherent in all seaplane configurations, as discussed later on under tail surface design. Therefore, it is advisable to keep the forebody length as short as passenger, cargo, and balance considerations permit with due allowance for a reasonable hull planform at the bow. A forebody length to beam ratio of 2.75 to 3.00 should provide an acceptable compromise.

Based upon these criteria, a 4000-lb seaplane hull 4.83 ft wide over the chines or spray rails at the step would have a 14.5 ft long forebody, a C_{Δ_o} value of 0.568, and a K value of 0.063 indicating satisfactory spray characteristics.

With the forebody length and beam established, the afterbody length and step depth may be determined by reference to Fig. 15-4. For a conventional hull, a deep step is preferable to a shallow one since the deeper offset in the hull bottom contour will provide better step ventilation—which results in improved lift-off, less critical trim limits, and reduced skipping and pitching tendencies. Additional ventilation may be realized by designing the beam at the step to be slightly narrower than that of the forebody maximum width ahead of the step. In other words, start bringing the hull in toward the sternpost ahead of the transverse step location. A step depth of 9 percent beam (at the chines) will be satisfactory for smaller seaplanes, reducing to 5 percent for hulls approaching 20,000 lb displacement.

To provide stable trim during planing, the hull step should be located on a line running approximately 10° aft from a keel perpendicular through the gross weight center of gravity (Fig. 15-1).

The afterbody sternpost and deadrise angles required for a stable hull are set forth in Figs. 15-5 and 15-6. These data are based upon actual water handling characteristics of many seaplanes as well as towing test results obtained from model flying boat hulls. As a practical consideration, afterbody bottom warping should not be necessary for hulls under 5000 lb displacement; the afterbody deadrise angle may be the same as that of the forebody at the step. Also, the af-

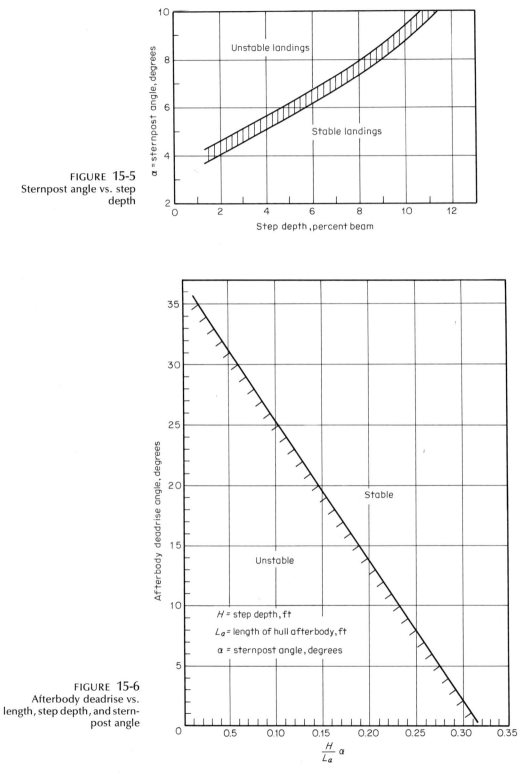

FIGURE **15-5**
Sternpost angle vs. step depth

FIGURE **15-6**
Afterbody deadrise vs. length, step depth, and sternpost angle

terbody angle must be adequate to keep the sternpost area clear of the water at the normal maximum hull trim angle τ of about 8°. A dragging afterbody or sternpost extends takeoff run at best, and may prevent takeoff of marginal performance seaplanes.

For planing tail hulls the afterbody should be slightly longer than the forebody. Due to the tail arm required for flight stability, this relationship develops quite naturally when the tail surfaces are mounted at the sternpost per modern amphibian design practice.

In closing this outline discussion covering conventional hulls, some realistic values of sea-state capability will be of practical use. There is a great tendency to expect too much of any hull, or to underestimate wave height, with equally unfortunate results in either case. Service reports indicate the following wave height capabilities may be *safely* anticipated, subject to superimposed swell conditions:

Hull displacement, lb	Wave height, ft
2,000	1
4,000	1.75 − 2
8,000	2.5
20,000	3.5
60,000	5
100,000	6

Since waves have been known to break over the flight deck of aircraft carriers, the use of amphibians and seaplanes for open ocean operation will be restricted to large displacement hulls and acceptable surface conditions. However, the sea-state capability of any flying boat hull may be considerably increased by the addition of a retractable hydrofoil extending from the keel, a new development briefly reviewed in the following section.

2. *Next Generation Hulls—Advanced Hull Design* A series of hydro-ski and hydrofoil development programs was sponsored by the Naval Air Systems Command and conducted by the author through 1970. This work resulted in the successful flight evaluation of a single, surface-piercing hydrofoil mounted on a conventional hull as an effective means of permitting a seaplane to operate in sea-state conditions beyond its normal capability. A comprehensive review of this program as well as hydrofoil seaplane detail design data are presented in Ref. 15.2.

Carrying this development further, it is possible to combine the hydrofoil with a faired step hull as a means of reducing cruise drag. Continued improvement through rounded or faired chines in combination with a faired step or a stepless hull design would reduce flight drag to a probable minimum for an amphibian. Such an advance must be thoroughly investigated by towing basin tests to prove the ability of the hull to unstick at hydrofoil unporting speeds and then to trim properly when planing on the foil.

The actual hydrofoil area required to support a 4000-lb displacement amphibian at 30 knots may be determined through the following typical calculation:

$W = \Delta_o = 4000$ lb

$S_W = 260$ sq ft (slotted flaps)

Hull trim angle prior to unporting hydrofoil $= 7.5°$

Wing $C_L = 1.9$ (wing angle of attack is sum of wing incidence angle plus hull trim angle)

Hydrofoil unporting speed $= 30$ knots

Note that the unporting speed must be high enough to provide effective aerodynamic control. If the hull becomes foilborne at too low a speed, the seaplane becomes uncontrollable due to lack of control surface effectiveness.

(i) Wing lift at 30 knots $= 0.0034\ C_L S_W\ (V_{knots})^2$
$$= 0.0034 \times 1.9 \times 260 \times (30)^2$$
$$L_W = 1510\ \text{lb}$$

(ii) Hydrofoil lift required $= \Delta_o -$ wing lift
$$L_H = 4000 - 1510 = 2490\ \text{lb}$$

(iii) $S_H =$ hydrofoil area; sq ft $= \dfrac{L_H}{1.424\ C_L\ \rho_{W_H}(V_{knots})^2}$

where $L_H =$ hydrofoil lift, lb

$C_{L_H} =$ hydrofoil lift coef. $= 0.27$ per Ref. 15.2, page 30

(Note: this 0.27 value is about maximum for a supercavitating hydrofoil. Foil angle of attack for this lift coefficient is 7.5°, the same as hull trim angle. This foil is therefore set at 0° incidence to keel.)

$\rho_w =$ water density $= 1.94$ for freshwater
$$= 1.99\ \text{for saltwater}$$

$V =$ hydrofoil unporting speed, knots

For this example: $S_H = \dfrac{2490}{1.424 \times 0.27 \times 1.94 \times (30)^2} = 3.71$ sq ft of hydrofoil planform area

The density of freshwater is generally used with this equation to assure adequate foil area for freshwater operation. The relatively small area required to provide 2490 lb of support at 30 knots is due to the fact that freshwater is about

(a)

(b)

FIGURE 15-7
The HRV-1 test amphibian
(a) planing on a supercavitating hydrofoil, and (b) taking
off with hydrofoil extended.
Howard Levy Photo

815 times the density of standard-day sea-level atmosphere, which in turn explains why the hull bottom and support structure must be very rugged.

Detail design data for a suitable hydrofoil section, support strut, and position along the hull bottom are covered in Ref. 15.2 and will not be discussed further in this book. However, it is interesting to note that a basic hull capable of handling 2-ft waves would be expected to operate safely in 4-ft seas when supported upon a hydrofoil extended 2 ft below the keel. The first successful working example of this concept, the HRV-1 hydrofoil amphibian, is shown planing on a single supercavitating foil in Fig. 15-7a. Flight photo Fig. 15-7b was taken during climb-out immediately after takeoff, offering a comparative view of required hydrofoil size and shape.

WING DESIGN   The wing and any related lift-increasing devices such as flaps or drooping ailerons should be designed to provide the minimum landing speed economically practical. The ability to approach safely at 60 knots and touch down at 45 knots opens many small bodies of water and small fields for possible landing sites. Further, at low stalling speed the water takeoff time and run will be considerably reduced compared with a design which must accelerate to a higher breakout speed. And reduced water contact time and speed mean safer operation in rough sea conditions.

After takeoff, climb becomes critical. Wing area and horsepower provide climb, but high aspect ratio wings have comparatively low induced drag at low (climb) speed and so offer increased ROC for a given wing area and power loading as covered at length in Chap. 6. In view of this, wing aspect ratios of 8 to 10 should be considered; the slightly greater weight of a high aspect ratio wing structure is more than compensated by increased performance throughout the flight envelope.

The two parameters of wing area and aspect ratio establish the wing planform geometry, which should include a slotted or similar type of lift flap to provide an acceptable lift coefficient. High drag devices such as split or plain flaps should be avoided since amphibians have more than enough drag for steep approaches. Flaps will generally improve water takeoff performance and should be designed to remain as clear of solid water as possible. Water impingement increases takeoff drag, extending takeoff run and time. Repeated spray contact can also damage light structures, so flaps must be designed with consideration for water operational requirements.

The wing airfoil should be selected to provide minimum aerodynamic moment and drag. Hull shape and thrust line location alone usually require considerable power "off" longitudinal trim capability to meet FAA low-speed stability criteria, without having added amounts of trim required by the airfoil section. This is particularly true at approach speeds in the landing configuration with flaps and gear down, power "off." Since selection of suitable airfoils has been covered in Chap. 6, this information will not be repeated here.

A design relationship peculiar to the amphibian occurs in the selection of wing angle of incidence relative to the hull keel. As mentioned under hull design, the keel should be inclined up and forward 1 or 2° at the step to broaden trim limits. When planing on the step, the hull will trim with the keel inclined about 8° to the water surface. Since maximum lift should occur with the hull at this attitude, wing incidence will be established accordingly, depending upon airfoil characteristics and takeoff flap deflection. Note that land-operating attitude requirements will have to be made compatible with the water-operating criteria; this occasionally becomes a challenging compromise because the afterbody sternpost must remain just clear of the water surface but also miss the airport runway during full stall landings. Because the wing incidence angle established for water lift-off will usually be too high for normal cruising flight trim attitude, amphibians tend to fly nose low during cruise. This improves visibility but requires some familiarization for level flight trim. The use of flaps tends to shift maximum lift to a lower angle of attack which, fortunately, reduces wing incidence and so improves hull trim during flaps retracted cruising flight.

Ailerons should be designed for minimum adverse yaw since flying boat directional stability tends to be marginal at best, as discussed in the following section. The ailerons must also be capable of lifting a low wing at a speed of about 20 knots in order to permit crosswind takeoff from water. Larger than normal ailerons are necessary to satisfy this requirement, with the usual result that they feel stiff in cruise flight. This condition may be improved by using a relatively thin (9 to 10 percent) airfoil section ahead of the aileron area. As with flaps, water impingement must be considered as an aileron design condition.

If specifications permit, wing area should be based upon 0.9 C_{Lmax} to permit future growth with acceptable takeoff speeds. For comparative assistance during preliminary design, average values of wing loading variation with gross weight have been plotted for single and multiengine amphibians in Fig. 15-8.

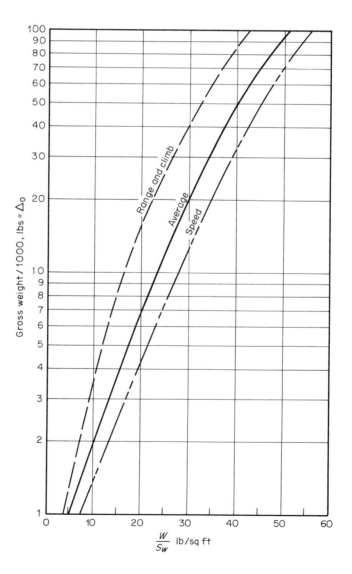

FIGURE 15-8
Wing loading vs. gross weight, based on 46 single- and multi-engine seaplanes

TAIL SURFACE DESIGN The T-tail arrangement with the stabilizer and elevator positioned at the top of the fin provides excellent spray clearance for the horizontal surfaces. With single-engine amphibians of this design, the horizontal tail may be located in the propeller slipstream to obtain minimum trim change with different power settings. In this position the horizontal tail also remains clear of flow changes due to flaps, although buffet warning may be quite light at the stall.

With the stabilizer acting as an end plate atop the fin, the vertical surface is more effective for a given area. Although this improves directional stability, fin design loads and the horizontal tail unsymmetrical loading will be slightly greater due to horizontal and vertical surface interaction in lateral gust and maneuvering conditions as discussed in Chap. 7. However, any increased directional effectiveness is very important since hull-type amphibian configurations tend to be marginally stable directionally due to the bow area ahead of the center of gravity. Not only does the hull profile move the center of lateral area close to the center of gravity, but as shown by Fig. 15-9 the bow shape will produce a lift component in yawed flight that tends to hold the amphibian in a displaced, yawed attitude during a skid, sideslip, or lateral gust (turbulent) condition. This inherent hull design problem may not be generally recognized, but it does exist and must be considered during preliminary design.

For seaplane design the horizontal tail must be able to trim large moment changes caused by a relatively high thrust line acting about the center of gravity, hull drag, step drag and location below the center of gravity, and flap application—all of which cause tail-load variations with speed and power

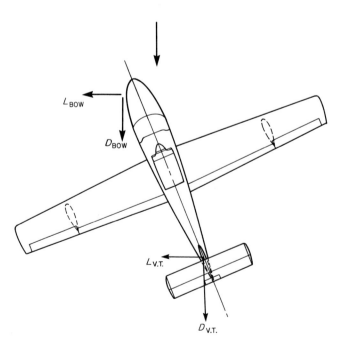

FIGURE **15-9**
Bow lift in yawed flight condition

changes. For larger amphibians an adjustable stabilizer is recommended, although the drive and pivot attachment must be ruggedly designed to withstand rough-sea operation, stall buffet, and possibly propeller slipstream.

Nose-up longitudinal trim with power "off" and flaps down may be critical at the forward center of gravity gross weight condition. If so, elevator tab effectiveness will be increased by static balancing the elevators or by adding a balance spring into the elevator control system; either method will overcome the natural tendency of an unbalanced elevator to trim trailing edge down in flight. Necessary design criteria for adequate and effective tail area will be found in Chap. 7.

LANDING GEAR DESIGN
The type of landing gear, conventional or tricycle, and main wheel location are important basic considerations for amphibian landing gear design. The final configuration will require many compromises because the landing gear must:

1. Have structural support adequate to carry land and water landing loads, although the heavily loaded hull or wing structure must be cut away to accept the retracted gear

2. Provide land landing clearance for the afterbody hull sternpost, yet the main gear support strut length must not cause major structural design problems

3. Be capable of land operation without nosing over when braked at full throttle (conventional gear) or tail squatting when empty (tricycle gear)

4. Be stowed or raised clear for water operation, but be immediately available for beaching or ramp approach

Amphibian pilots operating in bush country and remote areas prefer conventional (main wheels forward plus tailwheel) landing gear because it is easier to handle when leaving or entering the water and provides better rough-field operation than tricycle gear. These advantages have been covered in Chaps. 6 and 14 and illustrated in Fig. 6-1.

Despite the recommendation of conventional gear for utility operation, high-speed amphibians intended primarily for the executive and sports markets should be equipped with tricycle gear for easier handling on land. This type of amphibian will experience less frequent water operation than the utility models and should be designed with more consideration for land service.

If at all possible, the basic amphibian structural layout should be developed to accept both conventional and tricycle landing gear installations. Depending upon the intended design use, it should be possible to concentrate initially upon one landing gear system for early production and then develop the other type as the market expands.

The problem of landing on the water with the gear down and on land with the gear up cannot be overstated. Despite attempts to tie checklists into warning light systems through color coordination, the percentage of amphibian landings with the gear in the wrong position remains high. Using an amber light and amber checklist background for water landings (gear-up) and a green light and green checklist for land operation (gear-down) should help, but only if cockpit procedure properly includes reading the checklist prior to landing.

Some reduction in structural damage from land landings gear-up can be expected if a steel skeg or skid wear plate is fitted to the keel at the step, since the step area is designed to support such loads due to water pressure requirements.

Water landings gear-down generally cause the most damage and can be fatal unless the hull is extremely rugged. Some reduction in resulting repairs may be realized by designing the main strut attachment to withstand the torsional impact loading of a wheels-down water landing, but at a weight penalty that seems necessary for a safe and acceptable amphibian design.

POWERPLANT DESIGN   To provide required service, a utility amphibian must have sufficient thrust to assure rapid acceleration and climb-out at design gross weight on a 90°F day.

Rapid acceleration reduces water takeoff time and distance necessary to reach breakout speed—usually about 5 knots above stall for a reasonable margin of control. As previously mentioned, minimum water contact time is a safety feature serving to reduce wave impact exposure in rough-sea conditions as well as the possibility of striking floating or partially submerged debris. A gross weight climb rate of at least 800 ft/min under standard conditions is desirable to clear wooded shores, to overcome downdrafts, and to clear hills usually associated with rivers or lakes. The design objective is: reduced exposure to wave impacts on the water and to mountainsides when airborne.

To satisfy these criteria the design static thrust available should exceed a value of one-third gross weight. At an average static thrust of 4 lb/hp, a minimum of 330 hp would be required for a 4000-lb gross weight amphibian, representing a power loading of 12 lb/hp. Typical values of power loading vs. gross weight are plotted in Fig. 15-10 for actual single and multiengine flying boats and amphibians.

Regarding the powerplant package, it is quite obvious that the propeller or propellers must be located clear of all heavy spray and solid water thrown out by the hull. However, this requirement has been too frequently ignored in the past with the result that excessive propeller maintenance and replacement have created customer dissatisfaction with water operation in general. Protecting the propellers requires a raised powerplant installation whether the configuration is a single or multiengine design, while a high thrust line means fairly large trim changes with power. In view of this, three-bladed propellers should be used with larger engines to keep the thrust line as low as possible since they will deliver power from a smaller diameter than two-bladed types. The three-bladed propeller will also be quieter because of reduced diameter with lower tip speed at full power rpm.

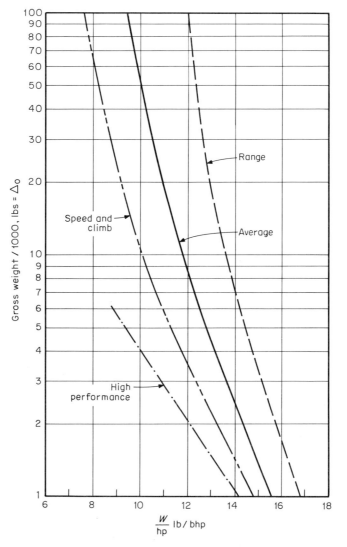

FIGURE **15-10**
Power loading vs. gross
weight, based on 46 sea-
planes

For larger amphibians a reversing pitch propeller should be provided for holding position in a seaway, backing down from a dock or beach, and braking dock approaches in adverse wind and current conditions. This feature is not mandatory for small amphibians since a paddle will usually provide adequate docking assistance.

To keep the single-engine amphibian propeller clear of spray requires either a pylon-mounted tractor or pusher nacelle located above the hull near the fore and aft center of gravity location; a tractor installation mounted ahead of the windshield over the bow; or a spray-rail-protected pusher assembly faired into the aft cabin section, which may experience considerable propeller erosion. For single-engine amphibians, the tractor installation will deliver more thrust in flight than the pusher design while also providing some hull lift if the wing center section or similar hull top contour is exposed to slipstream flow per the Teal amphibian of Fig. 4-13.

As an item of design interest, the one-engine-out directional control problems usually experienced with twin-engine designs can be virtually eliminated by using a tandem tractor-pusher installation. This package could be located above the hull centerline in a pylon-mounted "twin pack" nacelle.

If sound emission is to be reduced to levels necessary to satisfy current FAA environmental (noise) specifications, all new designs whether of single or multiengine configuration will require the lowest possible propeller tip speed combined with the use of effective mufflers.

TRANSVERSE STABILITY REQUIREMENTS

Flying-boat designers have used hull side sponsons, immersed wing tips, planing skis, and similar approaches to obtain transverse stability on the water. While such appendages are effective when underway, they must also provide static stability—and this requires adequate physical displacement.

For smaller amphibians, a wing float located near the tip is most satisfactory. To accommodate the weight of a person walking along the wing, which may occasionally be the only way to disembark, the tip float buoyancy should be sufficient to support a 200-lb load at the wing tip. This condition is usually critical for small flying boats, although the effect of wind side loads of 35 knots should be investigated for the design at rest in the water.

Although there are many complex formulas available to compute support float volume, for seaplanes under 8000 lb gross weight a suitable displacement value for each tip float will be obtained from the relationship

$$\Delta_{tf} = \frac{0.75W}{y} \quad \text{to} \quad \frac{1.25W}{y}$$

where Δ_{tf} = freshwater displacement of each tip float, lb

W = gross weight of amphibian, lb

y = distance from centerline to float center of buoyancy, ft

The tip float bottom surface should provide dynamic lift when running on the step to permit banking into the direction of turn without submerging the inside wing panel. By reducing the tendency to skid out of a turn, banking permits water turns of smaller radius and at higher speeds than would be possible with tip floats that could submerge underway.

Wing floats should be designed with a structural fuse attachment which will shear off when striking a log or similar object in the water, protecting the supporting wing structure from basic damage.

For high-performance amphibians, wing floats may be retracted in flight to act as end plates located outboard of the ailerons. But they must be designed to extend and retract at the same time to avoid directional stability problems.

OPERATIONAL
REQUIREMENTS

Although entire books have been written about seaplane handling and operational requirements, designers are mainly concerned with the safety and utility aspects of their particular configuration. A few design considerations not previously covered are:

1. Since an owner purchases an amphibian primarily because of its ability to operate from water, any design compromise should be resolved in favor of improved water handling characteristics, assuming pilot safety is not involved. This is a basic consideration for any amphibian design.

2. The importance of designing for the minimum possible landing speed cannot be overstated. As shown dramatically by Fig. 15-2, reducing the landing contact speed from 70 knots to 50 knots cuts the water impact load factor virtually in half. While this is not surprising since impact magnitude is a dynamic function of velocity squared, it is extremely significant from the safety standpoint. And not only during landing. Any reduction in gross weight stalling speed is accompanied by lower takeoff speed which in turn reduces takeoff run and time. Shorter takeoff run permits safe access to smaller bodies of water, while reducing water takeoff time minimizes contact exposure to rough seas and possible floating objects. Coupled with normally better hull characteristics at lower speeds, all factors associated with reduced landing speed act to increase water operational safety.

3. The amphibian is a utility airplane and hopefully a workhorse for serving remote areas. As a result, the ability to load and unload relatively dense cargoes such as packaged foods, chemicals, fuel drums, and sick people must be investigated and resolved during preliminary design.

4. Fuel quantity is another important consideration. Flight into remote areas usually means no refueling until return to home base. Depending upon the size of the amphibian as well as the operational specifications, fuel capacity for a minimum flight time of 5 hr should be provided, with optional capacity available for 8 to 10 hr flight at maximum-range cruise power. In addition to extending the radius of action to useful limits, such duration is a source of comfort during IFR conditions and will permit extended flights between established airports where fueling is possible. Fuel should be stored in wing tanks outboard of the hull; considerable structural relief will be realized at gross weight with full fuel in this location while passengers will be clear of fuel if structural damage occurs.

5. Docking, particularly in a seaway, strong current, or on a lee shore requires protected facilities of at least a minimum sort. A suitable dock for remote areas may be obtained by constructing a horseshoelike structure moored from the center of the closed end. This structure will stream with the open end downwind or down current, permitting the flying boat to taxi into a containing dock with off-loading ramps positioned either side of the hull as shown by Fig. 15-11. Such a mooring is almost mandatory for rocky and steep shores or beachless areas. It is unwise to leave any type of seaplane afloat at anchor by itself for overnight or longer; the hull plating could have

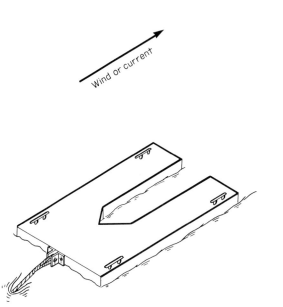

FIGURE **15-11**
Horseshoe dock design

been pierced by floating debris or a structural joint may have opened during the last landing. If the amphibian cannot be taxied ashore or beached for extended periods, it should be docked in such manner that it cannot sink. While automatic electric bailers may provide some mental relief for this condition, their starting float can stick or the aircraft battery may fail; at best, integral bailers are a secondary safety feature.

6. Mooring cleats located near the bow and tail are essential for water handling, mooring dockside, and anchoring. Each cleat and its supporting structure should be designed to withstand shear and tension loads separately applied (not combined) but equal to the amphibian gross weight. Although mooring line loads would never be expected to reach this value, an anchored amphibian will hook, snatch, dance, and generally work on the mooring line just like any boat, developing relatively large inertia load factors at the attachment cleat.

In addition, the normal wing-tie-down provisions should be included for securing on land.

7. Lastly, but as basic as any other design consideration, is the need to provide corrosion protection throughout the airframe. The amphibian must be able to operate off salt water with such minimum maintenance as washing with fresh water after each flight and regularly grease-filling rotating joints to keep out all water. This means alodining and priming all aluminum alloy parts; cadmium-plating steel pieces; sealing exposed skin joints, beam caps

and doublers, and major structural attachments; using aluminum alloy wheels and brakes and stainless steel control cables; and isolating dissimilar materials.

OPERATIONAL PROCEDURES

1. Most small seaplanes receive FAA approval for operation in 6 to 12-in. wave conditions (as measured from trough to crest). Although some aircraft may be safely flown off 24-in. high waves, operation of a small seaplane in seas of this size is not recommended procedure for anyone except the very experienced water pilot. In addition to the usual surface waves, the mouth of a sheltered bay opening onto the ocean will usually contain running swells which may combine with local wave conditions to provide extreme pitchup during a takeoff run into the advancing swells. Such surface conditions can result in becoming airborne prior to attaining flying speed and should be countered by running parallel to the wave front.

    It should also be noted that the wake of a powerboat can produce superimposed swells of very short wavelength which at the least result in a very uncomfortable takeoff run. Depending upon the size of the powerboat, such swells can also pitch a small seaplane into the air before reaching minimum flight speed.

    When experiencing steeper seas than anticipated during takeoff, abort the run by chopping throttle and turning crosswind (to the left or right, whichever direction is toward shore and calmer water). Because of the stabilizing action of the wing floats on midwing hull-type seaplanes, they will not capsize as would a pontoon seaplane caught in a high crosswind.

2. When judging surface conditions from approximately 500 ft above the water area intended for landing, it is suggested that the observer multiply the wave height estimate by a factor of 2. Water has a peculiar habit of growing in wave height the closer an airplane approaches the surface during final. As a general rule, large unprotected and open lakes are not suitable for water operation far from shore when surface winds exceed 10 mph. Although prior experience has shown that seaplanes can be full stall landed on water in winds up to 30 mph (the surface wind speed reduces the water contact speed by a like amount), under such wind conditions large lakes may develop very steep seas (short wavelength from crest to crest) that make takeoff quite dangerous—if not impossible.

    By contrast, when operating from the usual small lakes up to 1 × 2 mi in size, there is not sufficient distance (fetch) for large seas to develop; with these conditions, water flight may be safely conducted in winds up to 25 to 30 mph. Even so, the takeoff run can be uncomfortable.

3. Always plan your approach to land on the windward (direction the wind comes from) side or end of the lake or bay. Through this procedure, you will contact seas that have minimum fetch with correspondingly less wave height, as well as being near a protected shore if the sea state is considered too great for takeoff. Possible landing areas are shown on Fig. 15-12.

4. If caught in heavy seas after landing, do not attempt to take off. Taxi toward

the windward shore (the shore upwind of the airplane) at about 10 mph with hatches closed. If calmer seas are reached, takeoff can be made. It may also be possible to taxi into the lee of a point or behind an island or breakwater providing calmer sea conditions which permit takeoff (Fig. 15-12).

5. When it is not possible to find acceptable water conditions, taxi up onto a beach or moor the airplane on the *downwind* side of a dock or float located on the windward side of the lake or bay. After all, when on the water surface a seaplane is a boat and should be handled like one. So when mooring be sure to attach bow and stern lines as well as spring lines fore and aft to prevent airplane motion relative to the dock as shown in Fig. 15-13a. Leave sufficient slack to permit the airplane to float at least 1 ft away from the dock (the wind holds it away from the downwind side of a dock or float). If suitable line or docking cleats are not available, moor the plane with a single yoke line around both bow cleats and running to the downwind side of the dock; the surface wind will stream the seaplane away from the dock and the fin area will keep it headed into the wind weathervane fashion.

When no docking or beaching is possible, taxi close to the windward (upwind) shore, fasten a mooring line around the bow cleats to form a yoke leading to a single anchor line, and throw over the anchor. Numerous small boats will shortly appear to offer rescue—and may cause more problems than the wind conditions ever could.

6. As a final note, water flights should be conducted with all aboard wearing inflatable life vests, and with an anchor line, anchor, and paddle firmly se-

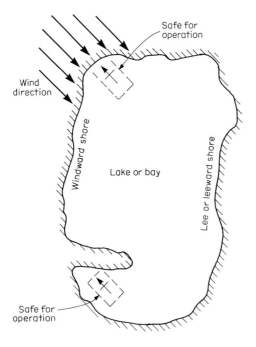

FIGURE **15-12**
Water areas for safe operation in heavy wind conditions

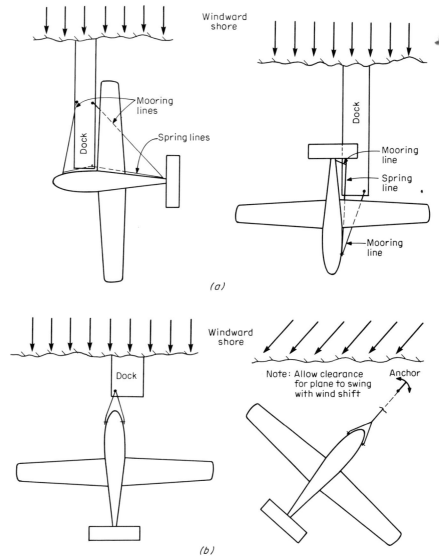

FIGURE 15-13
Possible docking and mooring conditions. (*a*) Taxi to dock and handle to positions shown. (*b*) Single-point mooring for short time.

cured in the cabin. For a more detailed and comprehensive review of water flying, Hank Kurt's interesting book is recommended per Ref. 15.3.

REFERENCES   15.1   W. C. Hugli, Jr., and W. C. Axt, *Hydrodynamic Investigation of a Series of Hull Models Suitable for Small Flying Boats and Amphibians.* NACA Technical Note 2503, November 1951. (Copies available from National Technical Information Service, Springfield, Virginia 22151.)

      15.2   D. B. Thurston and N. J. Vagianos, *Hydrofoil Seaplane Design.* Unclassified Final Report No. 6912 for Naval Air Systems Command under contract N00019-69-0475, May 1970.

      15.3   Franklin T. Kurt, *Water Flying.* The Macmillan Company, New York, 1974.

16 A LASTING FINISH

Many years ago, when the application of pure aluminum coatings to both surfaces of aluminum alloy sheet was first introduced as the "alclad" process, a shiny airplane was the height of fashion. When weather did not permit flights around and between airports, time could always be spent leisurely polishing the clad wings, fuselage, tail, and cowling while accepting compliments and discussing the latest airport news.

As spare time became ever more scarce and labor rates soared accordingly, the novelty of polished surfaces rapidly faded in favor of painted aircraft. This change in style was also fostered by the need to improve the visibility of increasing numbers of aircraft operating in clear sky or haze conditions, since an alclad finish tends to blend into its surroundings. It also became evident with time that (1) cladding would not stop corrosion over large scratched or abraded areas and (2) continued polishing would eventually wear away the protective surface—which in either case had to be painted or replaced.

So a whole field of protective processes and coatings evolved for aircraft use, with development and application rapidly accelerated by World War II camouflage requirements. Since shiny airplanes at altitude reflected sunlight like flying mirrors, they had to be given a lasting, nonreflective finish, with weather-resistant, fuelproof paints becoming the solution to this problem.

Although the basic corrosion protection process differs for steel and aluminum alloys, the same primers and finish paints may be applied to either material. Starting from the inside and working our way out to the surface, let us first briefly look at the different materials used for airframe construction and then see how they must be treated to provide the long life expected from modern aircraft.

AIRCRAFT MATERIALS   Figure 16-1 has been prepared to indicate the different aluminum and steel alloy strengths available and their accompanying Rockwell numbers. The Rockwell number is a calibrated reading obtained when a metal surface is pen-

Material	Ultimate tensile stress, psi	Material hardness	
		Rockwell B 100 kg	Rockwell C 150 kg
ALUMINUM ALLOYS			
5052-H34	34,000	B15/B22	
5052-H38	39,000	B28/B38	
6061-T4	30,000	B3/B15	
6061-T6	42,000	B38/B45	
2024 T3	Bare 64,000	B72/B76	
	Alclad 59,000	B67/B72	
7075-T6	77,000	B82/B87	
STEEL ALLOYS			
1015-1025 mild steel	55,000	B64/B68	
4130A (annealed)	70-85,000	B77/B87	
4130N (normalized)	90-95,000	B90/B94	
4130-heat treated	125,000		C26/C30
4130-heat treated	150,000 max.		C33/C37
4340-heat treated	180,000		C39/C42

FIGURE **16-1**
Materials and their hardness—aluminum and steel alloys commonly used in aircraft construction

etrated by a small steel ball or shaped diamond point loaded at a known value. The amount of penetration is an indication of material hardness which may be instantly read from a calibrated dial.

For example, the B-100 kilogram listing indicates a $^1/_{16}$ in. diameter size B hardened steel ball is pressed down upon a metal surface by a 100-kg force. If this ball penetrates the surface sufficiently far to produce the low reading of B20 on the indicator dial, we know from Fig. 16-1 that the material being checked is 5052-H34 aluminum alloy. In similar manner, if the penetration is only slight, say to a B85 reading, we know 7075-T6 aluminum alloy is being tested. Steel requires a sharper penetrator for the higher-strength alloys, so a Type C diamond point is used under a load of 150 kg (the right-hand column of Fig. 16-1).

Through this nondestructive inspection procedure it is possible to identify various materials and check completed items to ensure they are of the correct material, and to do so without harming the part. FAA inspection procedures require that every piece or lot of parts be checked in this manner for proper material strength, and also that certified records be maintained to document such inspection by airplane number. Through this process it is possible to determine which run of parts was used on each airplane if something should go wrong at a later date. Naturally, all this adds to aircraft cost, but also to flight safety. A more detailed review of nondestructing testing will be found in Ref. 16.1.

Aluminum alloy is the principal material used for aircraft construction for reasons indicated by Figs. 16-1 and 16-2. Figure 16-1 shows that 1015-1025 mild steel is not as strong as 2024-T3 bare or clad aluminum alloy sheet although it is 3 times as heavy per the weights given in Fig. 16-2. As a result, mild steel such as 1020 hot- or cold-rolled stock should not be used for aircraft parts because of its poor strength-to-weight ratio. By comparison, alclad 2024-T3 sheet offers about two-thirds the strength of normalized 4130 alloy steel at only one-third the weight.

Aluminum alloys are used in many forms: the most common being sheet made into skins, ribs, and frames, with extrusions used in quantity for beam caps, reinforcement angles, and small fittings tying parts together. Drawn tubing, rolled bar, special forgings, and castings are also fabricated from various alloys suitable for the different forming operations as listed on Fig. 16-3. For the highest strength-to-weight ratio aluminum alloy structures, bare sheet should be used because cladding material is of low strength, reducing allowable stress levels for clad sheet.

Extrusions cannot be clad because these shapes are formed by forcing aluminum which has been heated to a plastic state through a die of the desired cross section—much like making spaghetti. After being cut to length, each extruded piece is straightened by pulling on the ends the way string may be pulled tight (and straight).

Aluminum alloy tubing may be drawn or extruded. Drawn tubing is accurately formed by being worked to size through a series of forming dies. This type of tubing is seamless and stronger than extruded tubing, with the latter used primarily for noncritical secondary structures or nonaircraft commercial applications.

FIGURE **16-2**
Comparative properties of aircraft materials. (See Figure 16-3 for material specifications.)

Material →	2024-T3 Alclad sht.	2024-T4 extrusion	Normalized 4130 sheet		4130 H.T. to 125,000 psi
Thickness →	.010–.249	.050–.249	$\leqq.187$	$>.187$	As req'd
Ultimate tensile stress	59,000 psi	57,000 psi	95,000 psi	90,000 psi	125,000 psi
Yield tensile stress	39,000 psi	42,000 psi	75,000 psi	70,000 psi	103,000 psi
Compression yield stress	36,000 psi	38,000 psi	75,000 psi	70,000 psi	113,000 psi
Ultimate shear stress	38,000 psi	30,000 psi	55,000 psi	55,000 psi	82,000 psi
Ultimate bearing stress (2 dia. edge dist.)	114,000 psi	108,000 psi	200,000 psi	190,000 psi	251,000 psi
Weight, lb/cu in.	.10	.10	.30	.30	.30
e = elongation	12–15%	12%	10–15%	over 17%	10%

FIGURE 16-3
Material specifications for aluminum and steel alloys commonly used in aircraft construction

ALUMINUM ALLOYS

Type	Bare sheet and plate	Clad sheet and plate	Rolled or drawn bars, rods, and wire	Drawn tubing	Extruded bars, rods, tubing, and shapes	Forgings	Castings
2014	H.T. to -T6 AMS 4029	QQ-A-250/3	QQ-A-225/4		QQ-A-200/2	MIL-A-22771 Type 2014	
2024	QQ-A-250/4	QQ-A-250/5	QQ-A-225/6	WW-T-700/3	QQ-A-200/3		
5052	QQ-A-250/8		QQ-A-225/7	WW-T-700/4			
6061	QQ-A-250/11		QQ-A-225/8	WW-T-700/6 Hydraulic tubing MIL-T-7081	QQ-A-200/8	MIL-A-22771 Type 6061	
6063					QQ-A-200/9		
7075	QQ-A-250/12	QQ-A-250/13	QQ-A-225/9	WW-T-700/7	QQ-A-200/11	MIL-A-22771 Type 7075	
356							QQ-A-601 Cond. T6
Tenzaloy							MIL-A-12033 Class 2

STEEL ALLOYS

Type	Sheet, strip, and plate	Condition	Maximum strength	Seamless tubing	Bars and forgings
1015–1025	Commercial	Mild steel	55,000 psi	MIL-T-5066	Commercial
4130	MIL-S-18729	Normalized / Heat treat	<.187 thick—95,000 >.187 thick—90,000 / 150,000 psi max	MIL-T-6736	MIL-S-6758
4140		Heat treat	200,000 psi max	AMS 6381 AMS 6390	MIL-S-5626
4340	AMS 6359	Heat treat	260,000 psi max	AMS 6415	MIL-S-5000
321 & 347	MIL-S-6721	Stainless steel		MIL-T-8606	

Heat-treatable steels are used in aircraft construction when space and the superior fatigue qualities of steel are critical considerations; for example, wing attachment fittings that must be located as far apart between the top and bottom beam caps as possible, yet not protrude beyond the wing section contour.

Normalized 4130 steel is the alloy usually used for light aircraft steel parts and fittings, having a useful ultimate strength of around 90,000 psi as shown on Fig. 16-1. Fortunately, the strength of 4130 steel can be increased by mild heat treating to 125,000 psi; at which level it is twice as strong as 2024-T3 alclad, but still weighs 3 times as much. But any further increase in 4130 strength requires holding fixtures during heat treating, will result in parts and assemblies difficult to drill or rework, and may produce brittle structures.

The highest heat treat that should ever be considered for 4130 steel sheet, tube, or bar parts is 150,000 psi. At this value steel will be almost 3 times the strength of clad 2024-T3 at 3 times the weight. This trade-off is accompanied by much cost, with welded tubular assemblies that have been brought up to this strength running the risk of looking more like pretzels than their original shape.

When higher strength is required for large and thick machined parts, 4140 or 4340 steels should be used. 4140 alloy may be heat-treated to 180,000 psi and even to 200,000 psi for certain applications, if such strength is absolutely necessary; while 4340 can be taken to 260,000 psi. Although steel parts heat-treated to these stress levels will exceed aluminum alloy strength-to-weight ratios, the 4130, 4140, and 4340 alloys are difficult to machine even in the annealed or normalized state and must undergo expensive and time-consuming heat treatment and inspection procedures before being bolted into the airframe.

So it is not surprising that welded tubular and sheet steel structures have been replaced by aluminum alloy for production aircraft. Although this approach also eliminates highly skilled welding labor, it has resulted primarily from the fact that steel structures are basically heavy assemblies that can be justified only for specialized applications, such as the pilot protection and serviceability required for Ag plane operation shown by Fig. 4-5.

Figure 16-3 has been included to list various alloys and their uses as a procurement guide for readers interested in doing their own design or construction work. Whenever a specification is listed, the respective alloy is available in the form noted. When purchasing materials be sure to request documented certification of the alloy to specification, since papers supplying this quality assurance information will be required for FAA approval of any structural repairs or modifications. For example, when alclad 2024-T3 sheet is delivered, it should be accompanied by a notarized report stating the material was manufactured to the standards of specification QQ-A-250/5. But certifications will not normally be supplied unless requested, which should be done at time of purchase.

For the record, alloy sheet is available in the following thicknesses (in inches):

1. *Aluminum:* 0.016, 0.020, 0.025, 0.032, 0.040, 0.050, 0.063, 0.071, 0.080, 0.090, 0.125, and 0.190. Material of 0.250 thickness and above is available as plate.

2. *Steel:* 0.025, 0.032, 0.040, 0.050, 0.063, 0.072, 0.080, 0.090, 0.100, 0.109, 0.125, 0.160, and 0.187. Heavier material is available as plate, while 0.025-in. 4130 sheet is becoming very difficult to obtain.

Local metal supply sources should be contacted for additional information concerning aluminum and steel alloys. Free data are usually available, a typical example being the useful Alcoa handbook of Ref. 16.2.

Fiber glass, foam, and similar materials were briefly covered in Chap. 12 under the heading of composite structures. These synthetic materials have particular appeal for simplified homebuilt aircraft, having reached maximum application in the virtually 100 percent synthetic Rand KR-1 and KR-2 designs as well as Burt Rutan's VariEze of Fig. 3-5.

Sailplane designers are increasingly turning to fiber glass and foam construction, particularly for higher-performance models where a smooth exterior finish is of primary importance to drag reduction. Due to the absence of engine vibration, fiber glass structures stand up very well for this type of aircraft; although in off-field landing accidents glass offers much less protection than sheet aluminum.

If existing synthetic materials and construction techniques are used, powered aircraft designed for FAA approval will experience a structural weight penalty accompanied by considerable reduction in useful load. Hopefully, new composite materials and methods can be introduced to provide the same level of structural integrity and reproduction offered by sheet metal—and at similar cost.

In the meanwhile, it is possible to design an aluminum or steel alloy tubular fuselage structure capable of carrying primary loads and then enclose this assembly within a contoured envelope of fiber glass covered polyvinyl chloride foam. This method of construction permits integral fairing of wing fillets, canopy supports, tail surfaces, etc., as well as buried antennas—all within a polished, smooth skin which may be readily removed for access and inspection. Unfortunately, faired tubular construction is not suitable for working seaplanes because of water loads, floating debris, thin ice—which can act like flying slate to penetrate the covering—and the rigors of docking and loading. But it offers drag and assembly advantages for the next generation of high-performance, efficient landplanes—some of which are now being designed for metal and foam construction.

PROTECTIVE METAL
COATINGS

Typical of the compromises arising in most engineering work, the advantages offered by a metal structure are somewhat reduced by the threat of corrosion. Although aluminum alloy does not rust, it can surely corrode, particularly in the presence of a salt-laden atmosphere. And frequently aluminum corrosion is not evident until well advanced—similar to termites working in floor beams and sill plates. Steel is a bit more honest; corrosion is immediately evident as rust streaks leach out over everything nearby. Since neither form of corrosion can be tolerated for primary airframe components, methods have been developed to protect metal surfaces both before and after assembly.

Aluminum alloy sheets may be made with a clad coating on both faces and receive no further interior surface finish during the life of the airplane. Or more protection may be added in the form of a chromate conversion coating, a type of chromic acid chemical dip treatment generally referred to by the original trade name of "alodine" or "alodining." Bare aluminum should be alodined prior to painting, not only to protect the material but also to provide a better paint bond—which in turn further protects the material.

All aluminum alloy parts for seaplanes should be fabricated from clad sheet and then alodined before painting to provide maximum protection from salt-water corrosion.

Alodine touch-up kits are available for coating local scratches, abrasions, or surface repairs prior to finish painting and should be part of the maintenance inventory for every metal airplane.

Steel fittings and small sheet parts should be cadmium-plated to protect the metal surface from environmental corrosion. Cadmium plating is an electro-chemical process during which the exterior surface of steel parts is coated with a cadmium deposit varying from two ten-thousandths to five ten-thousandths (0.0002 to 0.0005) of an in. in thickness. A 0.0003-in. coating is standard in the lightplane industry and is referred to as Class 2, Type I Cadmium Plating to specification QQ-P-416. The cadmium surface formed is relatively soft and tends to corrode before steel will, thereby covering surface scratches and preventing steel from rusting.

Complex steel parts such as weld clusters and large assemblies should not be cad-plated. Chemicals used in plating may become trapped in recesses or weld laps and then slowly but surely eat away on the structure, accomplishing just the reverse of what was expected. To protect the surface, steel assemblies of this type should be sandblasted or wire-brushed and then immediately painted with primer.

When parts to be cad-plated have 180,000 psi or greater strength it is necessary to have them stress relief heat-treated prior to plating. This step is required to prevent surface embrittlement and possible development of fatigue cracks. Since small aircraft steel parts are rarely raised to such high stress levels, plating embrittlement usually is no problem. However, spring steel wire is of much greater strength and to prevent early failure must be properly stress-relieved when cad-plated, usually by a baking procedure.

Steel tubing weld assemblies require some form of internal protection to prevent moist air entrapped during welding from steadily corroding the structure from the inner to the outer surface. Unfortunately, many weld assemblies are not properly protected and subsequent repair becomes extremely costly, to say nothing of the potential danger of an in-flight structural failure occurring before corrosion is noted.

Although many commercial solvents are offered for steel protection, the best method for welded tubular assemblies seems to be a raw linseed oil internal rinse. To be effective, cluster joints must contain holes permitting internal flow of oil through the assembly, or be drilled in nonstructural areas for external draining. Once the assembly has been flushed of oil, any external drain holes must be sealed with plated steel drive screws (such as Air Force-Navy Standard Hardware listed in AN535). Boiled linseed oil or pure linseed oil is not necessary; a commercial grade of raw oil is equally effective and much cheaper. Some users recommend heating the oil to about 180°F to drive out moisture and to permit following oil flow through the structure by touching the exterior surface and feeling heated areas.

SURFACE FINISHES   Whether of aluminum or steel, surfaces exposed to the weather should be coated with primer paint after alodining or cad plating. Primer is a fast-drying finish that may be applied by brush, spray, or dip, leaving a zinc chromate or epoxy base for finish painting. Various types are available, with zinc chromate being somewhat lighter and cheaper than epoxy and, therefore, the type most commonly used for priming small aircraft.

If an untreated aluminum alloy surface is to be painted, a "wash primer" coating is recommended prior to spraying any final finish. Wash primer is superior to zinc chromate primer for this application because it tends to etch into bare and clad metal surfaces, providing an excellent bond for finish painting. Of course, all this doesn't come easily since wash primers are two-part compounds which tend to "string" or congeal if not properly mixed. But the added protection obtained from wash priming is worth the effort, although guidance from your local paint supply source is strongly recommended prior to application.

Scratches or gouges in a fiber glass structure should always be given immediate attention, because it is possible for water to work between the glass layers. This moisture may either expand as trapped steam when the airplane is parked in intense sunlight, or freeze and crack the structure locally during exposure to low ambient temperatures. Fiber glass repair kits are available for quick touch-up jobs and should be used as soon as possible after surface defects are noticed.

Automobile body filler will smooth metal and fiber glass surfaces when local dents or gouges cannot be worked out, and should be applied in just the same manner minor auto body repairs are made at home. A better and lighter material is a form of fiber glass filler known as Feather Fill, available from the source noted in Ref. 16.3. This filler may be sprayed or troweled in place on metal or fiber glass, sanded smooth, and then finish painted. The smoother the sanding job the better the surface; unfortunately, much elbow grease is essential to all good finish work.

Finish paints come in about as many brands, trade names, and different base compounds as there are colors. Although enamels and lacquers have been around for years, the enamels have recently branched into acrylic and polyurethane bases as well as the old standard type—each with its own group of enthusiasts. Polyurethane may be the easier to apply because of its fairly long pot life, but both acrylic and polyurethane enamels are two-part paints requiring an activator plus the use of stir pots during spraying. Either may be hot-sprayed for a fast drying finish and will provide excellent protection with a high gloss. So take your choice.

For fast backyard or hangar repairs, lacquer is easy to apply and offers adequate protection when the base coat is sound. Plain old 4-hr drying enamel will probably last longer, offer a higher gloss finish, and may be applied over any paint. Remember that lacquer cannot be applied over enamel; it will attack the surface and immediately develop a crinkle finish.

Although the choice of paint will probably depend upon the color match available, finish paint provides important protection for the airframe. Any scratches or local peeling should be covered as soon as possible, even if primer is the only finish readily available.

SEAPLANE PROTECTION

Seaplanes likely to operate in salt water must have special protection in addition to exterior paint. When possible all aluminum alloy parts should be fabricated from alclad sheet, alodined, and primed. Control cables should be of 7×19 stainless steel, specification MIL-C-5424, for maximum life and strength. All exterior cables should be coated with Par-Al-Ketone, and touched up as often as necessary to maintain a continuous coating.

All interior and exterior bolt heads, nuts, and bolt ends should also have a brush coating of Par-Al-Ketone, with interior bolts and nuts given an additional seal of Vaseline if located down low in bilge areas. This represents a lot of work, but so does replacement of badly corroded fittings. And stainless bolts or screws, contrary to popular belief, are not an acceptable substitute for cad-plated steel AN (Air Force-Navy Standard Hardware) bolts in aluminum alloy structures exposed to salt water. I learned this the hard way a few years ago when I finally had to chisel and drill out a steel bolt which in one sailing season had become frozen (corroded) into the extruded aluminum mast of my ocean racing sailboat. Galvanized dipplated steel parts really behave best around aluminum and salt water, but the plating is so soft and irregular in texture that galvanized bolts are not acceptable for joining highly stressed structural members.

With constant attention and a quick check of critical areas after each water flight, the special problems of saltwater operation should not demand excessive maintenance. Probably the most effective corrosion inhibitor of all is a thorough wash down with fresh water immediately after each briny dip. If washing is not possible, fly to a nearby freshwater lake or river and run through two or three landings (gear-up, if an amphibian); this is not as satisfactory as a good hosing, but much better than nothing.

*   *   *

NOTES AND REFERENCES

When all is said and done, keeping your airplane clean and polished contributes the most toward a lasting finish; and of course, the smoother the surface the lower the skin friction drag—and better the performance. A polished airplane also reflects pride of ownership.

So you personally gain all around if you can treat routine and finish maintenance the same way you would any pleasant hobby. The only way you can then improve upon caring for your airplane is by keeping it hangared. Being under cover is the ultimate way to increase the life of the entire airplane—in addition to assuring a lasting finish.

16.1   AC 43.13-1A, pp. 132–139.
16.2   *Alcoa Structural Handbook,* Aluminum Company of America, Pittsburgh, Pa.
16.3   Feather Fill is available in pint, quart, and gallon quantities from:
Fibre Glass-Evercoat Co., Inc.
6600 Cornell Road
Cincinnati, Ohio   45242

17 WHAT IS CERTIFICATION?

I'm sure we've all seen publicity releases concerning a proposed new airplane design, including a responsible first flight date plus the notation that FAA approval is expected about 6 months thereafter. And quite likely the person or management group responsible for each announcement was completely sincere and really expected certification by the date indicated.

The unfortunate part of all this being that aviation people tend to be overly optimistic, and they frequently have their best dreams shattered when aircraft development programs drag on and money drains away. This partially explains why so few new aircraft that do fly ever obtain production approval—with less than 10 percent of such projects ever receiving a TC, the Federal Aviation Administration Design Type Certificate.

After having lived through six type certifications with all their attendant delays, frustrations, and financial demands, I thought it might be helpful to pilots and budding entrepreneurs alike if we reviewed just what goes on during the certification process, including why so much time and money fly by between the first design sketches and final approval by the FAA.

Although most pilots are aware of the FAA requirements covering private aircraft design under Part 23 of the Federal Air Regulations (known as FAR Part 23), very few people fully realize the detailed work and procedures necessary to obtain a TC. And after reading this description of the process you may well wonder why any aeronautical engineer would go down the certification path a second time, to say nothing of five or six times.

THE MARKET ARENA   The general aviation industry is unique in that it tends to sell almost entirely to itself. That is, aircraft are designed, built, and sold to people interested in flying; powerplants and electronic equipment must be specifically developed for aircraft use; personal aircraft airports and fixed-base operations develop and expand only in communities having at least one enthusiastic and organized aviation group; and all this work originates with people personally interested in and familiar with the benefits of flying. As a result, the various segments of general aviation are closely interrelated; when one section experiences a reduction in sales volume, the entire industry soon follows, fanning the fire.

Since private aircraft and related flight training are really luxury items for the majority of owners and users, our industry rapidly responds to negative factors in the economy. When national income falls or major business profits decline, the resulting belt-tightening process rapidly squeezes out private flying. All of this means designers of small aircraft must be both attentive and responsive to current and developing industry trends, for these are potential customer requirements which will influence the acceptance of any new product.

DEVELOPMENT
OF DESIGN
REQUIREMENTS

A new aircraft design usually originates through the following procedures, either singly or in combination:

1. An extension or improvement of an existing design—such as increased wing area and useful load, or increased horsepower, cruising speed, and useful load.

2. In response to dealer requests arising from customer inquiry and/or operating experience. Hopefully, backed by dealer orders for the new model.

3. Through designer concept of a new type of configuration for which it is believed a market exists. This approach can be dangerously costly unless backed by firm orders from others willing to share the development risk and investment required.

4. A special configuration designed to order. This can be an experimental designer's ideal situation, as he or she will be paid for developing a new airplane. Since production runs are unlikely for special models, activity of this type is usually restricted to individual designers or very small aircraft companies.

Major details of any new airplane are first roughly sketched to support equally rough weight, balance, and performance estimates. Once the resulting concept receives customer or management approval, preliminary design work is started in exacting detail.

THE CONCEPT PHASE

From the basic preliminary sketches developed to launch the new project the design engineer next evolves the final airplane configuration, meanwhile balancing and juggling allowable weight items, budgeted development and production costs, tooling vs. labor-hour production trade-offs, and FAR Part 23 detail design requirements.

One of the first layouts will be an inboard profile, or sideview, showing the location of all systems and major structural members. This drawing requires considerable effort since it must realistically position passengers, equipment size and location, control system components and movement, instrument and radio location and space requirements, landing gear position and correct size of supporting structural members, tail clearance for maximum lift when landing, major structural bulkheads and longeron sizes, powerplant installation, pilot and passenger vision angles, location of passenger and service access doors, etc. Such basic requirements as FAA ground clearance for the propeller as well as safety belt and shoulder harness attachments must also be determined. Each of these items requires detailed design study and usually some compromise discussion before the design is frozen in final form.

All components must be checked for proper weight and center-of-gravity location at normal load, and also for maximum forward and aft center-of-gravity travel with disposable items in place and out to make sure the balance range falls within acceptable aerodynamic and ground handling limits. Of course, while all this is going on, company personnel feel obliged to comment on each line set down, even though such work may be very preliminary. Fortunately, many favorable ideas result from random comments, so all suggestions should receive studied attention—which time has shown to be the safer and more socially acceptable, though difficult, approach.

When the weight, balance range, structural arrangement, and aircraft outline appear acceptable, the design is considered fairly well set, subject to at least an informal mock-up of the cabin arrangement. For some reason, location of the seats, control wheel or stick, and rudder pedals never seems to transfer properly from paper to three dimensions, so some rearrangement of these items is usually required.

Once the configuration is approved, the decision to produce detail design drawings follows. At this stage, an aircraft company becomes committed to expenditure of a six-figure sum if the design is to be taken through FAA certification; a decision frequently made, incidentally, without any guarantee of sufficient production to realize at least the return of development capital.

For this reason, normal business financing sources are understandably reluctant to provide working capital unless the company has more than adequate collateral to cover any loans required. This means that if a small company does not have adequate net worth or a large cash flow developed by other product lines, it must find sources of risk capital before undertaking a new design. And if the new airplane is not successful, all responsibility falls upon the designer's shoulders.

Assuming the necessary financing is eventually obtained, we then proceed along the TC approval road by advising the FAA of the new program. Parallel with detail layout and design drawings, the engineering department must develop the applied loads which actually establish structural requirements. Part 23 specifications set forth these loading conditions in great detail, and for some items, such as tail surfaces, it may be necessary to calculate three or four design conditions before the critical set of loads is determined. Depending upon the size of the airplane, and whether or not it is subject to water loads and/or twin-engine operation, as many as 100 pages of applied loads data may be necessary. Of course, these end calculations may be backed by another 200 pages of detail analysis; so considerable engineering effort is required to prepare this basic information.

The applied loads are subject to close review and must be approved by the FAA since the safety of the resulting design depends upon the thoroughness and accuracy of these calculations. As a result, considerable time may pass before the applied loads are accepted. During this period the prototype (first experimental) airplane and tooling are growing at a fast pace, and any change in load requirements can be both time-consuming and costly. Again, the designer feels the impact of any modifications required.

DESIGN AND CONSTRUCTION OF THE PROTOTYPE

As detail tooling for the new design is completed, one or two sets of parts are run off and prototype assembly begins. If most of the structure is to be substantiated by detail mathematical analysis with a minimum of static test work, usually only one set of parts need be made; if extensive static test work is required by engineering, additional parts and assemblies must be fabricated.

Some designers prefer detail structural analysis to static tests of the basic components, since the margins of safety obtained through such analyses are available for future airframe growth and development. Also, complex static test programs may require so many calculations and test fixtures that this route proves more costly than detail analysis. However, the FAA does require certain mandatory static tests such as

· Wing panel assembly

· Wing torsional stiffness

· Typical wing rib

· Flutter investigation

· Control system

· Control surfaces

· Landing gear drop test

· Fuel tank endurance tests (and oil tank, if used)

· Pilot and passenger seats and belt attachments

· Baggage compartment floor

If the prototype airplane is to be used for these demonstrations, tests will be conducted at the design limit load condition which is approximately two-thirds of ultimate load. Any items tested to ultimate load (which includes a factor of safety of 1.50 times the design load value as discussed in Chap. 2) may not be flown and must be scrapped. Further, any future increases in gross weight or loading of any component proven by static test alone will require retest. So the load analysis better be correct when tests are run or they will have to be rerun, incurring a further loss of time and money.

As a matter of policy, all drawings, design analyses, and static test work are subject to detail review by FAA regional engineering personnel. For some items the FAA will accept design approval submitted by DER consultants. The DER is a Designated Engineering Representative appointed by the FAA as competent to perform engineering work on its behalf. A DER may be approved for airframe stress analysis, flutter analysis, systems and equipment design, powerplant design, flight test, or flight data analysis, and is of great assistance in preparing and expediting engineering work for TC design acceptance. However, even with DER approval in hand, FAA engineering reserves the right to review and disapprove any and all items, although such occurrences are extremely rare.

During the fabrication of all airframe parts and prototype assembly, company inspection precedures and records must be established and maintained in detail to substantiate the quality of work completed. FAA inspection will be provided by the assigned EMDO (Engineering and Manufacturing District Office) representative who should be brought into the project at the earliest possible time. It is most difficult to inspect detail components of a fully assembled wing, fuselage, or tail section; and even more difficult to dismantle assemblies to permit FAA inspection. For without FAA inspection approval the prototype cannot be flown by FAA test pilots, and therefore cannot be certified.

When the prototype is completely assembled, certain static tests must be performed and FAA approved prior to initial flight test; these include the control system and surfaces, wing torsion, and flutter investigations.

Following these demonstrations and supported by drawings and parts manufactured to EMDO inspection approval, the new airplane will be given an experimental airworthiness license permitting flight by company pilots. It should be noted that all detail drawings and structural analyses need not and probably will not have received FAA approval at this time, since flight work normally indicates a need for some minor or major changes.

At this point the FAA regional office may elect to convene a Preliminary Type Board for review of the new design. The purpose of this meeting, held at the manufacturing plant, is to submit the prototype to close study by all regional department heads. Following this review the board will prepare a list of recommended detail modifications considered necessary to meet FAA specifications. This is an important step in the TC procedure and indicates the program is moving along very well, regardless of the length of the resulting modification list.

FLIGHT TEST   The experimental prototype will be expected to remain in the vicinity of the home airport for the first 10 hr of flight. Test investigations must be conducted over fairly unpopulated areas (presumably to prevent dropping out of the sky onto housing developments), and no persons may be carried except those required to obtain flight data.

During company conducted flight tests, information will be gathered to substantiate compliance with Part 23 detail flight requirements. The FAA has a flight test form most useful in completing this work, which must cover stability throughout the flight envelope, design dives beyond the red line limit to be used for production aircraft, fuel system operation, engine cooling, ground (and water) handling, climb rate, and spins—to name the major areas presented in Ref. 17.1. Different interpretations of FAA flight requirements as well as variations in pilot opinion regarding handling characteristics can make the flight test period extremely frustrating at times. However, once company pilots agree that the airplane configuration meets all FAA flight requirements as understood, the design is considered frozen for TC approval.

Following that decision, the entire design data package must be submitted and approved before an FAA test pilot will be permitted to evaluate the airplane. Upon completion of regional engineering design approval, the FAA issues a Type Inspection Authorization (TIA) requiring the EMDO inspector to review the prototype for conformity with the approved drawings. This covers detail compliance with all drawings and manufacturing procedures, including any flight required modifications such as area changes, fillets, dorsal fins, revised cowling, equipment changes, etc. When this inspection is satisfactorily completed and documented, FAA flight test evaluation may begin. This level of achievement is a major step toward certification and one realized by a very small percentage of new design starts.

The FAA regional test pilot staff will assign three or more pilots to evaluate company flight data, with one person designated as project pilot. Assuming company evaluation has been conscientiously conducted, this final phase of certification should run smoothly. Unfortunately, differences in interpretation of test requirements and/or results can arise, frequently prolonging flight approval. If structural or equipment changes result from FAA flight evaluation, another TIA must be issued and another EMDO inspection completed before FAA flight work can be resumed. So this phase of evaluation can run into a series of 2- to 4-month delays for a small and seemingly uncomplicated design, as well as becoming very expensive with flight test work costing close to $100/hr and salaries to be paid as time goes by.

In addition, if an initial run of production aircraft is coming down the line, any delays in final TC approval can tie up large sums of scarce capital, while all flight required changes must be retrofitted to aircraft in process. For obvious reasons, a small company is well advised to start production *after* receiving FAA Type Certificate approval, and not before.

PHASES OF
FAA APPROVAL

Once the design TC has been granted, the company holding the TC has 6 months in which to establish effective manufacturing and inspection procedures for the new model, as demonstrated by essentially identical empty weights and flight characteristics between successive production aircraft. FAA inspection personnel will assist in defining critical items and required procedures. At the end of this period, a new aircraft company will be surveyed by the EMDO staff and any deficiencies noted. Inspection and manufacturing process procedures must then be documented by the company in accordance with a mutually acceptable format. Following approval of the resulting procedures manual, a new company is granted an APIS (Approved Production Inspection System) certificate, while an established airframe manufacturer will have the existing Production Certificate (PC) amended to include the new airplane. And this is all part of the design TC procedure; since without an APIS or PC in process or approved, it is not possible to sell production units of the newly TC'd design. After an APIS is granted, the manufacturer must continue to refine fabrication and inspection procedures leading toward PC status as soon as possible.

In summary, the five critical steps for obtaining production approval of a new airplane design are

1. Notifying the FAA regional office that a new design program will be initiated. The FAA office then assigns a program code to the proposed design. General arrangement drawings, as well as preliminary weight and balance plus performance data should be submitted at the time FAA consideration of the new design is requested.

2. Preliminary Type Board review.

3. Type Inspection Authorization, which represents completion of the engineering data review and approval of the aircraft design for conformity inspection prior to FAA flight evaluation.

4. The award of the design TC.

5. Development of manufacturing and inspection procedures covering the new design to assure consistent quality of production aircraft, resulting in the award of APIS or PC certificates.

These five steps normally require 24 to 36 months of concentrated effort, accompanied by considerable engineering ability and much capital. So you can see why new aircraft are safe, but expensive.

SOME CLOSING
THOUGHTS

Although I do not agree with all Part 23 requirements, and particularly object to preparing fail-safe design and fatigue analyses for small aircraft used only a few hundred hours a year, there can be no argument that current FAA flight specifications produce a very stable, easily handled airplane. As flight proficiency requirements were lowered over the years to accommodate students and occasional pilots holding only a basic private license, aircraft design standards were raised to provide forgiving wings. And this is a proper philosophy—flying should not be a constant test of pilot skill if engineering can make it otherwise. The airplane should be developed to serve the average pilot, not the pilot to serve the airplane by virtue of superior skill.

On a constructive note, every designer has the opportunity to study and incorporate ways to eliminate pilot error items before the airplane flies. Based upon over 35 years of design experience, I believe many so-called pilot error incidents are more the fault of the designer than of the pilot. In fact every aircraft designer has an implied responsibility to eliminate pilot error at the source by "designing out" opportunities for error. Fortunately we are realizing this level of improvement to an increasing degree in today's aircraft, but still have a way to go as every pilot knows. It is my sincere hope this book has made a positive contribution toward achieving that goal, while also adding to every pilot's understanding of how his or her airplane is designed and why it flies as it does.

REFERENCES   17.1   (a) Flight Test Report Guide; FAA Form 8110-11 prepared by the FAA.
(b) "Engineering Flight Test Guide for Small Airplanes"; Directive 8110.7 prepared by the FAA.
Contact any FAA regional office for copies of the above guides.

INDEX